# SUPERCOOLING

Edited by **Peter Wilson**

**Supercooling**
Edited by Peter Wilson

**Published by InTech**
Janeza Trdine 9, 51000 Rijeka, Croatia



**Publishing Process Manager** Sasa Leporic

**Technical Editor** Teodora Smiljanic

**Cover Designe**r InTech Design Team

Additional hard copies can be obtained from orders@intechopen.com

Supercooling, Edited by Peter Wilson
p. cm.
ISBN 978-953-51-0113-0

For EU product safety concerns:
IN TECH d.o.o., Prolaz Marije Krucifikse Kozulić 3, 51000 Rijeka, Croatia,
info@intechopen.com or visit our website at intechopen.com.

# Meet the editor

Professor Wilson gained his undergraduate degree in physics and then postgraduate degrees in electronics and in optics from Otago University in New Zealand. His interests then changed to soft condensed matter and in particular the antifreeze proteins polar fish use to survive in ice-laden waters. Postdoctoral work included measuring the thickness of the ice-water interface and for much of his career Professor Wilson has been studying the nucleation of supercooled solutions and how best to measure the nucleation temperature. In 2005, Roskilde University, Denmark, awarded him the degree Doctor Scientiarum. Professor Wilson now works in the Faculty of Life and Environmental Science at the University of Tsukuba in Japan.

# Contents

# Preface

The metastable state of liquids which are supercooled has been the subject of interest for over 300 years. Supercooled water and aqueous solutions are found in the atmosphere, in cold-hardy organisms, and in many industrial applications. For example, freeze tolerance and the associated avoidance of supercooling, is the dominant cold-related strategy in Southern Hemisphere insects, being found in 77% of cold hardy insects. In industry, thermodynamic systems such as cool boxes have a significant financial driving force for supercooling to be controlled and this is also covered in this book.

One less well known aspect of supercooling is during the process of cooling sugar-based (non-ionic) surfactants used as plasticizers, and the associated glass transition. Such transitions are of interest in the food, cosmetics, and pharmaceutical industries.

In the optics industry the interface stability of fluoride compounds has significant consequences. In fact melt growth of any single or mixed fluoride systems possesses substantial advantages over growth from solution because it may provide conditions for growing homogeneous crystals with two–three orders of magnitude higher crystallization rate. Melt growth is the only method appropriate for developing effective techniques for production of crystals with high optical quality

In metallurgy the process of crystallization in a system far from equilibrium has features which manifest themselves in the morphology, crystal growth velocity, and segregation of dissolved alloy components. Simulations of dendritic growth and coarsening for Al-Si alloy during isothermal solidification are carried out by phase field method are also discussed in this book.

**Peter Wilson**
Professor on the Global 30 Program at the Faculty of Life and Environmental Sciences,
University of Tsukuba
Japan

# 1

# Supercooling of Water

Peter Wilson
*Tsukuba University*
*Japan*

## 1. Introduction

Water has a complex phase diagram with more than 16 crystalline phases, two glass phases and liquid which displays many unique behaviors, especially in the region of -45 °C (Mishima and Stanley 1998, Stokely et al. 2010). It can remain a liquid even under conditions where a more stable phase exists, and in those conditions is said to be supercooled. Supercooled water can be prompted to turn into ice by seeding, and the best seed is ice itself. The nucleation of supercooled solutions has been studied for almost 300 years beginning with the early works by Fahrenheit (see Shaw et al. 2005). Water has long been known as an unusual liquid and significant research continues today into anomalies such as the compressibility (Abascal and Vega 2011) and the fascinating tendencies when water is confined to nanoscale dimensions (Strekalova et al. 2011). High pressure density fluctuations are also an ongoing research area (Mishima 2010), as is the so-called "no man's land", a region of the phase diagram existing below the homogeneous nucleation temperature (~ -38 °C) and above where amorphous ice can be found (~ -118 °C) (Moore and Molinero 2010).

Research involving supercooled water encompasses many fields of science and atmospheric research comprises a large fraction of these works (DeMott 1990, 1995). Ice formation in the atmosphere affects rainfall and snowfall as well as the level of solar radiation reaching the Earth's surface and so is very topical of late (Sastry 2005, Hegg and Baker 2009). In clouds, supercooled water droplets are thought to sometimes freeze homogeneously through the organization of water molecules into an ice lattice without the need for any external seeding agent (Tabazadeh et al. 2002, Ansmann et al. 2005). Picolitre volumes of liquid water can be cooled in the laboratory to about –39 °C before freezing occurs and this homogeneous nucleation temperature is usually denoted as $T_{homo}$ or $T_{hom}$ (Koop and Zobrist 2009, Stan et al. 2009).

In biological systems some plant cells adapt to subfreezing temperatures by deep supercooling, with temperatures as low as –60 °C reported (Kasuga et al. 2007, 2010). In the case of insects it has been shown that most have the ability to supercool in the absence of gut content, but this does not automatically mean that they cold-hardy (Bale and Hayward 2010, Doucet et al. 2009). Supercooling to –40 °C has been reported in some freeze-avoiding, cold-hardy insects (Sformo et al. 2011). Conversely freeze tolerant insects, such as the New Zealand alpine weta, avoid supercooling by causing ice to form at high sub-melting point temperatures (Wharton 2011).

The addition of solutes to water causes a depression of $T_{hom}$ (Koop 2004) but in contrast the addition of ice nucleating agents (INAs) raises $T_{hom}$, and as we will see later also $T_{het}$, the heterogeneous nucleation temperature.

The usage, or avoidance, of supercooled water can be important in industry. For instance, thermal storage air conditioning systems use water as a phase change material and supercooling is a serious problem. Controlling the phenomenon is highly advantageous (see for instance Chen et al. 1999).

Since supercooled water is a metastable state, most studies on supercooled water actually are concerned with the liquid to solid transition (Langham and Mason 1958). In order to describe the liquid to solid transition mathematically Classical Nucleation Theory (CNT) was formulated by Frenkel and others in the 1930s. The history of CNT is outlined in a thorough review by Kathmann (2006).

Concerning nomenclature, the temperature at which a solution spontaneously freezes when cooled below its equilibrium freezing temperature, $T_f$, is denoted variously as the "kinetic freezing point" (MacKenzie, 1977), the "temperature of crystallization" (Vali 1985) and the "nucleation temperature" (Kristiansen et al. 1999), regardless of whether it is $T_{hom}$ or $T_{het}$. For biological solutions, or even for whole organisms, this temperature of spontaneous freezing is also often called the "supercooling point" (SCP) (Zachariassen 1985) and this is the notation adopted in this article.

## 2. Homogeneous nucleation of supercooled water

The homogeneous nucleation temperature $T_{hom}$ of liquid water is generally accepted to be approximately -39 °C. Below this temperature even finely dispersed water droplets will freeze instantly. It has also been proposed that this temperature is somewhat volume dependent and this concept, together with the reported anomalous density data of supercooled water, are discussed by Hare and Sorensen (1987). A brief review of the homogeneous nucleation of liquids can also be found in Debenedetti and Stanley (2003).

In practice, the phenomenon of homogeneous nucleation is actually very difficult to achieve under laboratory conditions. In order to approach the limit of homogeneous nucleation, much care must be taken in the preparation of the sample to avoid impurities, which may lead to heterogeneous nucleation. One method to achieve homogeneous nucleation takes advantage of extremely small volumes of ultra-pure water, which are immersed within oil emulsions (Broto and Clausse 1976). As the sample volume becomes smaller, there is a greater probability that impurities are absent from the liquid droplet. The oil acts to prevent extra surfaces or sites on which the sample may nucleate. A second method to achieve homogeneous nucleation involves levitating a small volume of aqueous solution by using an electromagnetic field (Kramer et al. 1999). Although this elegant method eliminates the use of a container, which again may introduce unwanted sites for nucleation, it does not ensure a sample free of impurities. Hence, most, if not all, nucleation encountered in the laboratory and in practical experience is heterogeneous. Even nucleation of clouds in the atmosphere is now believed by some to occur entirely or mostly on a seed particle such as ammonium sulfate (Hung et al. 2003).

Since any supercooled biological system is not actually composed of ultra-pure water, the effect of solutes on the SCP must be addressed, even though solute-induced decrease of $T_{hom}$ is somewhat difficult to measure. Rasmussen and McKenzie (1972) reported the $T_{hom}$ for a variety of aqueous solutions of increasing concentration, and their data suggested that $T_{hom}$ decreased by about twice the equivalent melting point depression, for a given volume and measurement technique. This ratio is usually denoted $\lambda$. More recently, Koop et al. (2000) analyzed the reported $T_{hom}$ of 18 different solutes as a function of solute molarity. In essence their work showed the ratio is unity for all physiological strengths, and that the identity of solute species is unimportant. Miyata et al. (2002) argue that there exists a strong correlation between $T_{hom}$ and ionic radius of alkali ions and/or halide ions. They argue that some of the results of Koop et al. were at best an approximation, at least in the case of ions. Koop et al. (2004) subsequently wrote a more thorough review where they also discuss some desirable future experiments.

In another review Zachariassen and Kristiansen (2000) discuss homogeneous nucleation, and argue that it may actually be the mechanism of freezing in some biological systems. They contend that nucleation in some insect species is not triggered by ice nucleation agents of any sort and that they instead undergo homogeneous nucleation. Wilson et al. (2003) found their argument flawed due to its reliance on the interpretation of the results of Bigg (1953). The "homogeneous" nucleation results of Bigg were brought into question by Langham and Mason (1958) from the same laboratory a few years later, and it seems that the results of Bigg have never been reproduced. The high $T_{hom}$ values reported by Bigg are possibly due to some heterogeneous process involving a reaction at the water-oil-surfactant interface.

## 3. Heterogeneous nucleation of supercooled water

Heterogeneous nucleation and reports surrounding it have been reviewed by Sear (2007) who points out that if we concede to having a meager understanding of homogeneous nucleation then our understanding of heterogeneous is even worse. Heterogeneous freezing, where the freezing event is initiated on a foreign surface, particle or even molecule, has been hypothesized to occur in four different ways: deposition nucleation (Dymarska et al. 2006, Kanji and Abbatt 2006, Kanji et al. 2008, Mohler et al. 2006); condensation freezing (Diehl et al. 201); immersion freezing (Marcolli et al. 2007) and contact freezing (Cooper 1974, Durant and Shaw 2005, Fukuta 1975a,b, Pruppacher and Klett 1997, Fornea et al. 2009). This range of freezing mechanisms makes experimental determination of $T_{het}$ problematic. For instance, it has been found that freezing seems to occur at higher temperatures in contact mode than immersion mode (Pruppacher and Klett 1997). Similarly it has been suggested that even with homogeneous nucleation the initial event is occurring at the water surface as opposed to within the bulk (Tabazedah et al. 2002) and evidence for this has also been provided by Shaw et al. (2005).

Apart from ultra-pure water sequestered in emulsions to reduce the contact with solid surfaces, all other aqueous solutions will undergo heterogeneous nucleation (Fletcher 1969). The liquid sample must be housed in a container of some form and even so-called "pure" water will, in general, have some impurities about which nucleation might proceed. From a free energy point of view, it is more favorable to grow an ice embryo on a two-dimensional surface than in a three-dimensional surface-free volume of water (Duft and Leisner 2004, Sigurbjornsson and Signorell 2008).

When the volumes of water used in an experiment are larger than the micron-sized droplets found in emulsions, and they are supercooled, $T_{het}$ varies markedly due to varying amounts of impurity particles, which may act as sites where nucleation may occur. Hosler and Hosler (1955) used a variety of sizes of capillary tubes and found that, even when the capillaries had an diameter of only 0.2 mm, the lowest temperature they could reach with water samples was -33 °C, at which point heterogeneous nucleation occurred. Most workers who use differential scanning calorimeters (DSC) use water as a control at one time or another. These workers usually find that the typical sample volume comprised of only 5 µl of pure water will invariably freeze in the DSC pan at temperatures ranging between about -21 °C and -25 °C (Wilson et al. 1999), which is far from $T_{hom}$.

With slightly larger volumes, studies have shown that 200 µL of clean, reagent grade distilled water sealed in a glass NMR tube typically freezes at temperatures around -14 °C, and that this temperature is somewhat container dependent when there is no efficient nucleator present in the aqueous sample (Heneghan et al. 2002). Using even larger volumes tends to make it more difficult to achieve low nucleation temperatures, simply because it enhances the probability of the presence of an efficient nucleator. Dorsey (1948) published a very thorough study cleanliness, glass type, and the effects of water conductivity on the heterogeneous nucleation temperature of water. Sample sizes of approximately 4 mL were used and at no time was he able to cool the water below about -19 °C before heterogeneous nucleation occurred. Inada et al. (2001) have managed to supercool several hundred mL of water down to temperatures around -12 °C, a major achievement for such a large volume of water.

## 4. Measuring the supercooling point

The stochastic nature of the value of $T_{het}$ is not always realized or well defined historically (Barlow and Haymet 1995, Heneghan et al. 2001). Often the SCP is measured by sealing the solution into a small capillary, and decreasing the temperature of the capillary linearly as a function of time at some preset rate until the solution freezes. This process is then usually only repeated a few times, employing different samples from the stock solution in each successive run. This procedure misses one of the most important aspects of this phenomenon, namely the inherent width of the distribution of SCP values (Vali 2008, Wharton et al. 2004). In the case of whole animals, the procedure is essentially the same, but great care is taken to ensure that the animal is not seeded with ice, which would prematurely induce freezing in supercooled fluids (Ramlov 2000). In this method, the lowest temperature reached prior to the sample freezing is defined as the SCP of the solution. Freeze-tolerant animals may survive the experience, and the same sample may be used several times to investigate the natural width of this supercooling point temperature, but not so with freeze avoiding animals.

Quite often, SCP determinations have been made from only a handful of measurements on each sample, or on each stock solution or on each group of animals, and the resulting values quoted with standard a deviations calculated in the usual way. However, if so few data points are determined, the likelihood of measuring the most probable nucleation temperature is small. In fact, Haymet and co-workers have shown that up to 200-300 measurements are needed on a single sample to determine accurately the nucleation temperature. They use an automatic lag time apparatus (ALTA), to study the statistics of liquid-to-crystal nucleation (Heneghan et al. 2004, Barlow and Haymet 1995). The machine repeatedly cools, nucleates and thaws a single, unchanging sample of solution. It operates in a linear cooling mode, in which the temperature of a single sample of liquid is decreased by a constant rate until the sample freezes. The set-up and operation of the apparatus for inorganic aqueous solutions has been described in Heneghan et al. (2002), Wilson et al. (2003, 2009, 2010), Wilson and Haymet (2009). Other workers are now using a similar analysis (see for example Seeley et al. 1999).

A typical data set from ALTA is shown here. Known as a Manhattan, it gives the time each run spends supercooled, and since cooling is linear this is also the temperature of each freezing event.

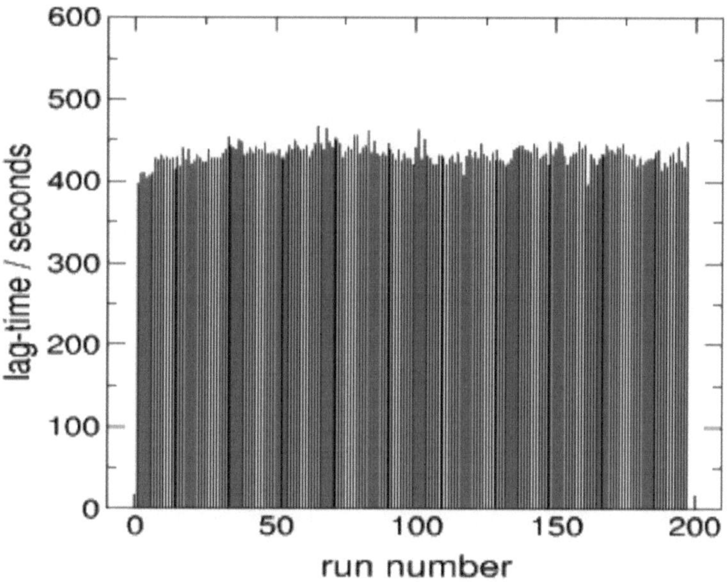


Fig. 1. Manhattan for a typical set of runs on ALTA, showing the stochastic nature of nucleation, where each run on the same sample freezes at a different temperature.

When data from ALTA is analyzed further it can produce a survival curve, a type of cumulative curve, showing the spread of freezing temperatures, as shown in fig 2.

The curves shown in Figure 3 show nucleation of the same single water sample in the same tube both with, and without, a crystal of silver iodide added to lower the free energy barrier. The SCP is then defined as the 50 % height of the survival curves, and has been shifted warmer by 7.7 C with the addition of the silver iodide.

Classical nucleation theory (CNT) was developed to describe the vapor to liquid transition (see for example Auer and Frenkel 2001) and is generally not appropriate for the liquid to solid transition. Results from ALTA experiments with pure water and water with an added catalyst have shown that CNT produces values incorrect by many orders of magnitude (Heneghan et al. 2001). The ALTA results show that the size of the critical ice nucleus needed to initiate the phase transition is much smaller than CNT predicts. Also, these

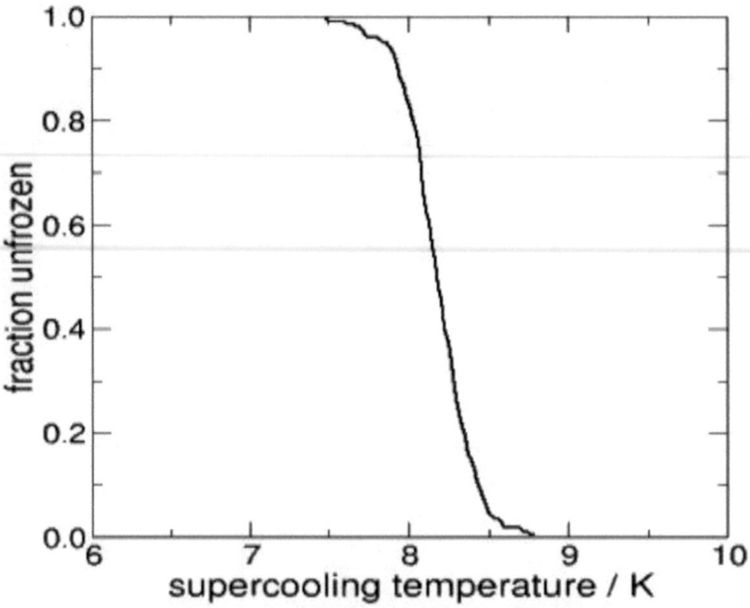


Fig. 2. Nucleation survival curve for an ALTA sample set showing the spread of nucleation temperatures. One of the most useful measures from such a curve is the 10-90 width, in this case about 0.7 °C.

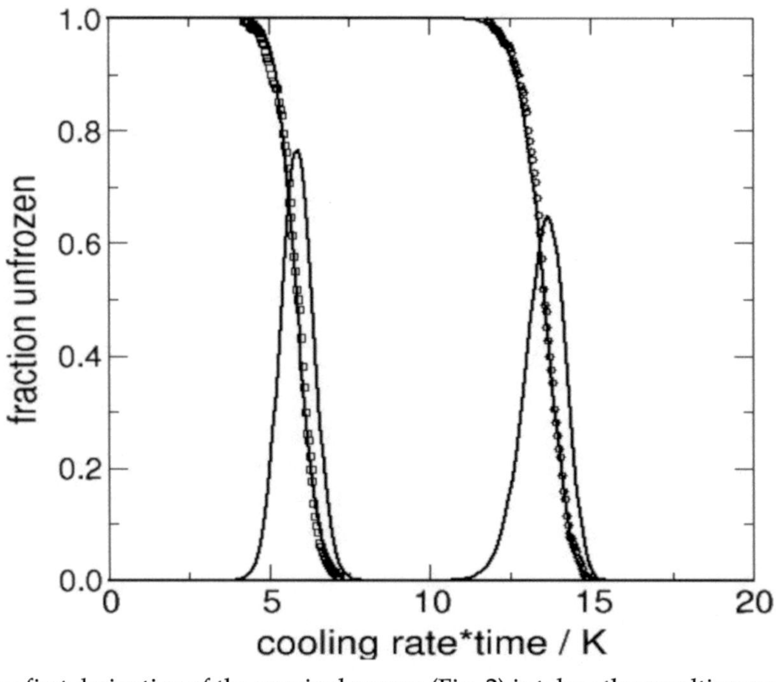


Fig. 3. If the first derivative of the survival curves (Fig. 2) is taken the resulting peaks show the probability of nucleation as a function of temperature. The 10-90 width is almost exactly the same as the full width at half height of these peaks.

experimental results are in agreement with recent theoretical calculations of Oxtoby who has used density functional theory to show that CNT is overly simplistic (see for example Shen and Oxtoby 1996).

The natural definition of the SCP is the temperature at which the survival curve crosses the 50 % unfrozen mark, namely the temperature at which on average half of the samples are frozen and half of the samples are unfrozen. For the data shown in Fig. 2, the proposed SCP is 8.17 K below the melting point. However, this survival curve also provides natural error bars for the SCP. By measuring the 10-90 width (the range of temperature where the sample is 90% unfrozen to the temperature where the sample is 10% unfrozen), an upper and lower bounds emerge naturally from this analysis. Here the 10-90 width is 0.7 K. This large spread in the temperature of nucleation for the exact same sample demonstrates further the point that many repetitions are needed. In other experiments utilizing droplets with volumes of 1 μl, some 500 to 600 thermal cycles have produced a 10-90 width of 0.75 C (Seeley et al. 1999).

## 5. Effect of electric field on supercooled water and nucleation

The ability to promote or suppress nucleation would have profound impacts on many fields such as cryopreservation, prevention of freezing of crops, cloud seeding, snow making to name but a few. One possibility for such manipulation might be the addition of an electric field. The effects of an electric field however are still under debate. Wilson et al. (2009) could find no effect on $T_{het}$ for DC fields up to $10^5$ $Vm^{-1}$, and similarly Stan et al. (2011) found no effect on $T_{hom}$ for fields up to $1.6*10^5$ $Vm^{-1}$. In contrast, Wei et al. (2008) found that fields of up to $1*10^5$ V/m could affect the SCP, albeit by only 1.6 °C. It cannot be ruled out that the inherent stochastic spread of $T_{het}$ would show such a spread in any event. In another study Ehre et al. (2010) found that water freezes differently in positively and negatively charged surfaces of pyroelectric materials.

## 6. Effects of solutes on the nucleation of supercooled water

With regards the SCP one important question is whether, for a given solution and container, added solutes actually decrease $T_{het}$ by an amount which is the same as the melting point (m.p.) depression, or twice as much, or three times? Even this seemingly simple measurement has proved somewhat problematic. Koop et al. (2000) claim that the identity of the solute species has no effect whatsoever on the ratio of the SCP depression to the m.p. depression. In a similar way the correlation between ionic radius and the SCP ratio has also been studied (Miyata et al. 2002). Block and Young (1979) reported that added glycerol decreases $T_{het}$ of some particular solutions by more than three times the equivalent melting point depression. Wang and Haymet (1998) have shown that even within the simple sugars, the amount of supercooling decrease for a given volume differs from one isomer to another. They found that trehalose and sucrose decreased $T_{het}$ in a modulated DSC further than glucose and fructose.

This ratio, $\lambda$, has also been examined closely by Duman et al. (1995) and $\lambda$ has been found to vary from unity to about two by Block (1991). It has been argued theoretically that $\lambda$ should be unity, even for homogeneous nucleation (Franks, 1981). In whole animal studies on species lacking ice nucleating agents, $\lambda$ has been reported to be closer to three (Somme 1967). In contrast, in all other studies where potent nucleators are present, $\lambda$ has been reported to have a value very close to unity (Lee, 1981). Zachariassen (1985) found that adding either saline or gylcerol at concentrations up to 2.5 osm. increased the SCP (same volume, same container) by a factor between 1.4 and 1.5. More recently Zachariassen and Kristiansen (2000) contended that the polyol accumulation in freeze-tolerant insects generally only decreases the SCP by a ratio of unity.

In the case of $T_{hom}$ Kanno et al. (2004) found that $\lambda$ is affected by the nature of the solute, contrary to the conclusion of Koop et al. (2000) and is close to 2. Kimizuka et al. (2008) found that $\lambda$ depends on the molecular weight of the species, for PEG, PVP and dextran, and values vary between 1.5 and 4.5. They also found that values correlate with the log of the self diffusion constant. Takehana et al. (2011) found that aqueous solutions of $H_2SO_4$ did not follow the linear relationship and that $\lambda$ was in fact a quadratic relationship with molarity.

It is clear that to measure accurately the nucleation temperature, many more measurements are needed than have typically been made in experiments published to date. Also, the solute dilution series must be carried out in the same container and under the same conditions, such as rate of cooling. Wilson and Haymet (2009) have investigated the effect of solute concentration on $T_{het}$ of aqueous solutions of both NaCl and D-glucose. Using the ALTA technique allowed the dependence of $T_{het}$ on solute concentration to be determined with statistical significance. The results showed that the solute-induced lowering of $T_{het}$ was in fact $\lambda$=2, at any fixed concentration, the same factor reported for homogeneous nucleation experiments with small molecular weight solutes.

## 7. Theories of nucleation

The tool used most often in modeling studies of liquid to solid nucleation is CNT. This theory uses the capillarity approximation whereby the properties of the critical cluster (and smaller) are considered equal to those of the bulk new phase. This approximation is questionable when the number of molecules making up the cluster is perhaps a few hundred to a few thousand. A clear description of the shortcomings of CNT can be found in Erdemir et al. (2009) who advocate a two-step model for nucleation of solids. A thorough review of theories of nucleation has been given by Hegg and Baker (2009), who also provide an overview of the state of the art with regard to theories of nucleation in the atmosphere.

Statistical mechanics has also been used to develop another theory of nucleation of this phase transition, the inadequacies of which are reviewed in Ford (2004). Classical density function theory (DFT) (Granasy 1999) has been recently used to look at heterogeneous crystal nucleation (Kahl and Lowen 2009) who describe it as an ideal tool to look at this problem. In contrast, for deep quenches spinodal nucleation theory is said to be needed (Wang and Gould 2007) who modeled the homogeneous and heterogeneous nucleation of Lennard Jones liquids. The concept of cluster chemical physics and associated dynamical nucleation theory (DNT) is discussed by Kathmann (2006). He has been able to do little more than point out the inadequacies however, since the constants used in the models are not currently known with sufficient accuracy. A look at CNT and quantum nucleation theory as it pertains to liquid to gas nucleation can be found in Maris (2006).

## 8. Supercooling in biological systems

The supercooling abilities, and otherwise, of some insect classes is reviewed by Doucet et al. (2009). Current climate change and the effects on cold-hardy insects has been discussed by Bale and Hayward (2010) who also include a brief overview of the supercooling abilities of over-wintering insects. The supercooling abilities of some plants, including trees, are discussed in Kasuga et al. (2007) and a general overview of nucleation and anti-nucleation in biological systems is given by Zachariassen and Kristiansen (2000).

The special case of ice-binding proteins and the effects on $T_{het}$ is examined more closely now. These special classes of proteins are often called antifreeze proteins (AFPs) and are thought to bind to ice to stop macroscopic growth inside many organisms. However, it is their effect on supercooled solutions which is of interest here.

Some of the body fluids of polar fishes, such as the eye, are supercooled for the duration of the life of the fish, albeit by less than 1°C. The gut contents are not supercooled since ice crystals will almost certainly be present as the fish swallow sea water. In the large Antarctic toothfish *Dissostichus mawsoni* there may be as much as one liter of blood, and if no ice were ever to enter through wounds or the gill filaments this may be supercooled for as long as 50 years. Even at 1 °C of supercooling such fluids still have the chance of heterogeneously nucleating ice and the presence of AFPs would then be necessary. It is still unclear whether one purpose of AFPs in polar fishes is to inhibit nucleation in the blood, since ice crystals are in fact present in the gut, and at times in the blood. Clearly the main job of AFPs is to inhibit the growth of crystals already present in environments conducive to growth.

Many insects/arthropods deliberately choose supercooling as a freeze-avoiding strategy and they too have AFPs in their haemolymph. In these cases the question remains as to whether the AFPs inhibit the nucleation of ice.

Conversely, biological ice nucleation has also been the focus of much research, especially in relation to plants and crop protection (Levin and Yankofsky 1983). It is well known that some bacteria produce very effective ice nucleation proteins (INPs) to enhance nucleation of ice at very high subzero temperatures. This topic is reviewed in Burke and Lindow (1990) who modeled these large proteins and assigned sizes and nucleation temperatures for particular scenarios. Since AFP molecules are thought to bind to ice to stop growth it is a small step to study larger proteins which bind water molecules to themselves in order to make a large enough "ice crystal" to pass the Gibbs free energy barrier and cause the solidification event. There have been sporadic reports of solutions with AFP being able to supercool further than workers would have expected, however, as we have seen there are inherent difficulties in measuring accurately the SCP. Duman (2002) has produced some interesting results with citrate. Basically, he found that AFPs from the beetle *Dendroides*, together with glycerol or citrate, can eliminate the activity of potent ice nucleators and thus lower the SCP further than without the added citrate. This concept seems not to have been explained fully to date. In contrast, Larese et al. (1996) found that adding AFP type I to animal cell suspensions actually increased the incidence of intracellular ice formation. The explanation being that the AFPs bound water to themselves and helped the embryonic crystal grow to sufficient size to cross the energy barrier.

The question arising from the work on AFPs and supercooling is as follows: Do AFPs find embryonic ice crystals and bind to them and stop them becoming large enough to cross the free energy barrier or do they bind to the most likely site of ice nucleation and mask it from water molecules, thus lessening the probability of nucleation? Wilson and Leader (1995) argued for the latter. Before any further insight can be gained into the effect of AFPs on nucleation it is necessary to be able to measure the supercooling point much more accurately in biological systems than has perhaps been generally possible to date. Holt (2003a,b) showed that antifreeze glycopeptides (AFGP) at 1% concentration could significantly lower the SCP when diluted with tap water. However when there was a strong nucleator present the SCP increased, and he put this down to the AFP joining the nucleators together, although there appeared to be no real evidence for that.

In a recent study Wilson et al (2009) examined the effects that antifreeze proteins have on the supercooling and ice-nucleating abilities of aqueous solutions. Using the ALTA technique, they showed several dilution series of Type I antifreeze proteins. Results indicated that, above a concentration of ~8 mg/ml, ice nucleation is enhanced rather than hindered. They went on to present a new hypothesis outlining three components of polar fish blood that which they believe affect its solution properties in certain situations.

## 9. Conclusion

All nucleation of supercooled biological solutions or whole animals is heterogeneous and this is probably true for most non-biological solutions as well. Many repetitions on the same or "identical" samples are required to measure accurately the SCP, and its partner quantity, the inherent width of the survival curve. Also, solutes decrease the SCP of solutions by twice as much as the equivalent melting point depression.

## 10. Acknowledgement

The author would like to thank Tony Haymet for many years of collaboration and helpful advice.

## 11. References

Abascal, J. L. F. and C. Vega. (2011). Note: Equation of state and compressibility of supercooled water: Simulations and experiment. J. Chem. Phys. 134, 186101-186102.

Ansmann, A., Mattis, I., Müller, D., Wandinger, U., Radlach, M., Althausen, D. and R. Damoah. (2005). Ice formation in Saharan dust over central Europe observed with temperature/humidity/aerosol Raman lidar, J. Geophys. Res., 110, D18S12.

Auer, S. and Frenkel, D. (2001). Suppression of crystal nucleation in polydisperse colloids due to increase of the surface free energy. Nature 413, 711-712.

Bale, J. S. and S. A. L. Hayward. (2010). Insect overwintering in a changing climate. J. Exp. Biol. 213, 980-994.

Barlow, T. W. and Haymet, A. D. J. (1995). ALTA: An automated lag-time apparatus for studying nucleation of supercooled liquids. Rev. Sci. Instrum. 66 (4) 2996-3007.

Bigg, E. K. (1953). The supercooling of water. *Proc. Phys. Soc. B* 66, 688- 694.

Block, W. (1991). To freeze or not to freeze? Invertebrate survival of sub-zero temperatures. Functional Ecology. 5, 264-290.

Block, W. and S.R Young. (1979). Measurement of supercooling in small arthropods and water droplets. Cryo-Letters, 1, 85–91.

Broto, F. and D. Clausse. (1976). A study of the freezing of supercooled water dispersed within emulsions by differential scanning calorimetry. *J. Phys. C: Solid State Phys.* 9, 4251- 4258.

Burke, M. J. and Lindow, S. E. (1990). Surface properties and size of the ice nucleation site in ice nucleation active bacteria: theoretical considerations. Cryobiology 27, 80-84.

Chen, S. L., Wang, P. P. and T. S. Lee. (1999). An experimental investigation of nucleation probability of supercooled water inside cylindrical capsules. Exp. Thermal and Fluid Sci. 18, 299-306.

Cooper, W. A. (1974). A possible mechanism for contact nucleation, J. Atmos. Sci., 31(7), 1832–1837.

Debenedetti, P. G. and H. E. Stanley. (2003). Supercooled and glassy water. Physics Today June, 40 - 46.

DeMott, P. J. (1990). An exploratory study of ice nucleation by soot aerosols, J. Appl. Meteorol., 29(10), 1072–1079.

DeMott, P. J. (1995). Quantitative descriptions of ice formation mechanisms of silver iodide-type aerosols, Atmos. Res., 38(1), 63–99.

DeVries, A. L. and Wilson, P.W. (1988). Ice in Antarctic fishes. Cryobiology. 25, 520-521.

Diehl, K. , Quick, C., Matthias-Maser, S., Mitra, S. K. and R. Jaenicke. (2001). The ice nucleating ability of pollen: Part I: Laboratory studies in deposition and condensation freezing modes. Atmospheric Research, 58, 2, 2001, 75-87.

Dorsey, N. E. (1948). The freezing of supercooled water. Trans. Am. Philos. Soc., 38, 248-256.

Doucet, D., V. K. Walker and W. Qin. (2009). The bugs that came in from the cold: molecular adaptations to low temperatures in insects. Cell. Mol. Life Sci. 66, 1404-1418.

Duft, D., and T. Leisner. (2004). Laboratory evidence for volume-dominated nucleation of ice in supercooled water microdroplets, Atmos. Chem. Phys., 4(7), 1997–2000.

Duman, J. G. (2002). The inhibition of ice nucleators by insect antifreeze proteins is enhanced by glycerol and citrate. J. Comp. Physiol. *B* 172, 163-168.

Duman, J. , Olsen, T., Yeung, K. and F Jerva. (1995). The roles of ice nucleation in invertebrates,.R.E Lee, L.V Gusta, Editors, Biological Ice Nucleation and Its Applications, Amer Phytopathological Society pp. 201–219.

Durant, A. J., and R. A. Shaw. (2005). Evaporation freezing by contact nucleation inside-out, Geophys. Res. Lett., 32, L20184,

Dymarska, M., B. J. Murray, L. Sun, M. L. Eastwood, D. A. Knopf, and A. K. Bertram (2006), Deposition ice nucleation on soot at temperatures relevant for the lower troposphere, J. Geophys. Res., 111, D04204,

Ehre, D., E. Lavert, M. Lahav and I. Lubomirsky. (2010). Water freezes differently on positively and negatively charged surfaces of pyroelectric materials. Science 327, 672-675.

Erdemir, D., A. Y. Lee and A. S. Myerson. (2008). Nucleation of crystals from solution: classical and two-step models. Accounts of Chemical Res.42, 5, 621-629.

Fletcher, N. H. (1969), Active sites and ice crystal nucleation, J. Atmos. Sci., 26(6), 1266–1271.

Franks, F. (1981). The nucleation of ice in undercooled aqueous solutions. Cryo-Letters 2, 27–31.

Ford, I. J. (2004). Statistical mechanics of nucleation: a review. Proc. Instn. Mech. Engrs. 218, C, J Mechanical Eng Sci. 883-899.

Fornea, A. P., Brooks, S. D., Dooley, J. B. and A. Saha. (2009). Heterogeneous freezing of ice on atmospheric aerosols containing ash, soot and soil. J. Geophys. Res. 114, D13201, (pp12).

Fukuta, N. (1975a). Comments on "A possible mechanism for contact nucleation", J. Atmos. Sci., *32*(12), 2371–2373.

Fukuta, N. (1975b). A study of the mechanism of contact ice nucleation, *J.* Atmos. Sci., *32*(8), 1597–1603.

Granasy, L. (1999). Cahn-Hilliard-type density functional calculations for homogeneous ice nucleation in undercooled water. J. Mol. Structure. 485, 523-536.

Hare, D. E. and C. M. Sorensen. (1987). The density of supercooled water. II. Bulk samples cooled to the homogeneous nucleation limit . J. Chem. Phys. 87, 4840 - 4846.

Hegg, D. A. and M. B. Baker (2009) Nucleation in the atmosphere. Rep. Prog. Phys. 72, 056801 (21pp).

Heneghan, A. F., Wilson, P. W. and A. D. J. Haymet. (2002). Statistics of heterogeneous nucleation of supercooled water, and the effect of an added catalyst. Proc. Natl. Acad. Sci. 99. 9631-9634.

Henaghan, A., Wilson, P. W., Wang, G. and A. D. J. Haymet. (2001). Liquid-to-Cystal Nucleation: Automated Lag-Time Apparatus to study supercooled liquids. J. Chem. Phys. 115, 7599.

Holt, C. B. (2003a). Substances which inhibit ice nucleation: a review. CryoLetters 24, 269-274.

Holt, C. B. (2003b). The effect of antifreeze proteins and poly(vinyl alcohol) on the nucleation of ice: a preliminary study. CryoLetters 24, 323-330.

Hosler, C. L. and C. R Hosler. (1955). An investigation of the freezing of water in capillaries. Trans. Am. Geophys. Union, 36, 126 - 132.

Inada, T., Zhang, X., Yabe, A. and Y. Kozawa. (2001). Active control of phase change from supercooled water to ice by ultrasonic vibration. 1. Control of freezing temperature. Int. J. Heat Mass Transfer, 44, 4523–4531.

Kahl, G. and H. Lowen. (2009). Classical density functional theory: an ideal tool to study heterogeneous crystals nucleation. J. Phys. Condens. Matteter 21, 464101 (7pp).

Kanji, Z. A., and J. P. D. Abbatt. (2006). Laboratory studies of ice formation via deposition mode nucleation onto mineral dust and n-hexane soot samples, *J. Geophys. Res., 111*, D16204,

Kanji, Z. A., Florea, O. and J. P. D. Abbatt. (2008). Ice formation via deposition nucleation on mineral dust and organics: Dependence of onset relative humidity on total particulate surface area, Environ. Res. Lett.,3(2), 025004.

Kanno, H., Miyata, K., Tomizawa, K. and H. Tanaka. (2004). Additivity rule holds in supercooling of aqueous solutions. J. Phys. Chem. A 108, 6079-6082.

Kay, J. E., Tsemekhman, V., Larson, B., Baker, M. and B. Swanson. (2003). Comment on evidence for surface-initiated homogenous nucleation. Atmos. Chem. Phys. Discuss., 3, 3361-3372.

Kasuga, J., Y. Fukushi, C. Kuwabara, D. Wang, A. Nishioka, E. Fujikawa, K. Arakawa and S. Fujikawa. (2010). Analysis of supercooling-facilitating (anti-ice nucleation) activity of flavonol glycosides. Cryobiology 60, 240-243.

Kasuga, J., K. Mizuno, K. Arakawa and S. Fujikawa. (2007). Anti-ice nucleation activity in xylem extracts from trees that contain deep supercooling xylem parenchyma cells. Cryobiology 55, 305-314.

Kathmann, S. M. (2006). Understanding the chemical physics of nucleation. Theor. Chem. Acc. 116, 169-182.

Kimizuka, N., Viriyarattanasak, C. and T. Suzuki. (2008). Ice nucleation and supercooling behavior of polymer aqueous solutions. Cryobiology 56, 80-87.

Koop, T., Luo, B., Tsias, A. and T. Peter. (2000). Water activity as the determinant for homogeneous ice nucleation in aqueous solutions. Nature 406, 611-614.

Koop, T. (2004). Homogeneous ice nucleation in water and aqueous solutions, Z. Phys. Chem., Suppl., *218*(11), 1231–1258.

Koop, T. and B. Zobrist. (2009). Parameterizations for ice nucleation in biological and atmospheric systems. Phys. Chem. Chem. Phys. 11, 10839-10850.

Krämer, B. H., O. Hübner, H. Vortisch, L. Wöste, T. Leisner, M. Schwell, E. Rühl, and H. Baumgärtel. (1999). Homogeneous nucleation rates of supercooled water measured in single levitated microdroplets, J. Chem. Phys., 111(14), 6521–6527.

Kristiansen, E., Pedersen, S., Ramlov, H. and K.E Zachariassen. (1999). Antifreeze activity in the cerabycid beetle rhagium inquisitor. J. Comp. Physiol. B, 169, 5–60.

Langham, E. J., and B. J. Mason. (1958). The heterogeneous and homogeneous nucleation of supercooled water, Proc. R. Soc. Lond. A, *247*(1251), 493–504.

Larese, A., Acker, J., Muldrew, K., Yang, H. and McGann, L. (1996)/ Antifreeze proteins induce intracellular nucleation. Cryo-Letters 17, 175-182.

Lee, R. E., Zachariassen, K. E. and J.G Baust. (1981). Effect of cryoprotectants on the activity of hemolymph nucleating agents in physical solutions. Cryobiology, 18, 511–514.

Levin, Z., and S. A. Yankofsky. (1983). Contact versus immersion freezing of freely suspended droplets by bacterial ice nuclei, J. Appl. Meteorol., 22(11), 1964–1966.

MacKenzie, A. R. (1997). Are the (Solid-Liquid) Kelvin equation and the theory of interfacial tension components commensurate? J. Phys. Chem. B 101, 1817-1823.

Marcolli, C., S. Gedamke, T. Peter, and B. Zobrist. (2007). Efficiency of immersion mode ice nucleation on surrogates of mineral dust, Atmos. Chem. Phys., 7(19), 5081–5091.

Maris, H. J. (2006). Introduction to the physics of nucleation. C.R. Physique 7, 946-958.

Mishima, O. (2010). Volume of supercooled water under pressure and the liquid-liquid critical point. J. Chem. Phys. 133, 144503-1- 5.

Mishima, O. and H. E Stanley. (1998). The relationship between liquid, supercooled and glassy water. Nature 396, 329-335.

Miyata, M., Kanno, H., Niino, T. and K. Tomizawa. (2002). Cationic and anionic effects on the homogeneous nucleation of ice in aqueous alkali halide solutions. Chem. Phys. Lett. 354, 1-2, 51-55.

Möhler, O., P. R. Field, P. Conolly, S. Benz, H. Saathoff, M. Schnaiter, R. Wagner, R. Cotton, and M. Kramer. (2006). Efficiency of the deposition mode ice nucleation on mineral dust particles, Atmos. Chem. Phys.,6, 3007–3021.

Moore, E. B. and V. Molinero. (2010). Ice crystallization in water's "no-man's land". J. Chem. Phys. 132, 244504-1-10.

Pruppacher, H. R., and J. D. Klett. (1997). *The Microphysics of Clouds and Precipitation*, 2nd rev. and enlarged ed., 954 pp., Kluwer Acad. Publishers, Dordrecht, Netherlands.

Ramlov, H. (2000). Aspects of natural cold tolerance in ectothermic animals. Human Reproduction 15, 26-46.

Rasmussen, D. H. and A. P. MacKenzie. (1972). in *Water Structure at the Water Polymer Interface,* edited by H. H. G. Jellinek (Plenum, New York), pp. 126.

Sastry, S. (2005). Ins and outs of ice nucleation. Nature 438, 746-747.

Sear, R. P. (2007). Nucleation: theory and applications to protein solutions and colloidal suspensions. J. Condens. Matter 19, 033101 (38pp).

Seeley, L. H., Seidler, G. T. and Dash, J. G. (1999). Apparatus for statistical studies of heterogeneous nucleation. Rev. Sci. Instrum. 70, 9, 3664-3667.

Sformo, T., J. McIntyre, K. R. Walters Jr., B. M. Barnes and J. Duman. (2011). Probability of freezing in the freeze-avoiding beetle larvae Cucujus clavipes puniceus (Coleoptera: Cucujidae) from interior Alaska. J. Insect Physiology 57, 1170-1177.

Shaw, R. A., A. J. Durant, and Y. Mi. (2005). Heterogeneous surface crystallization observed in undercooled water, J. Phys. Chem. B, 109(20), 9865 – 9868.

Shen, Y. C. and Oxtoby, D. (1996). Nucleation of Lennard-Jones fluids: a density functional approach. J. Chem. Phys. 105, 6517-6521.

Sigurbjörnsson, O. F., and R. Signorell. (2008). Volume versus surface nucleation in freezing aerosols, Phys. Rev. E, 77(5), 051601.

Somme, L. (1967). The effect of temperature and anoxia on the hemolymph composition and supercooling in three overwintering insects. J. Insect Physiol., 13, 805–814.

Stan, C. A. , G. F. Schneider, S. S. Shevkoplyas, M. Hashimoto, M. Ibanescu, B. J. Wiley and G. M. Whitesides. (2009). A microfluidic apparatus for the study of ice nucleation in supercooled water drops. Lab on a Chip. 9, 2293 - 2305.

Stan, C. A., S. K. Y. Tang, K. J. M.Bishop and G. M. Whitesides. (2011). Externally applied electric fields up to $1.6*10^5$ V/m do not affect the homogeneous nucleation of ice in supercooled water. J. Phys. Chem. B 115, 1089-1097.

Stokely, K. M. G. Mazza, H. E Stanley and G. Franzese. (2010). Effect of hydrogen bond cooperativity on the behavior of water. PNAS 107, 4, 1301-1306.

Strekalova, E. G., M. G. Mazza, H. E. Stanley and G. Franzese. (2011). Large decrease of fluctuations for supercooled water in hydrophobic nanoconfinement. Phys. Rev. let. 106, 145701-1- 145701-4.

Tabazadeh, A., Y. S. Djikaev, and Reiss. (2002). Surface crystallization of supercooled water in clouds, Proc. Natl. Acad. Sci., 99(25), 15,873–15,878.

Takehana, M., C. Viriyarattanasak, K. Kajiwara and H. Kanno. (2011). Non-existence of the linear relation between $T_H$ (homogeneous nucleation temperature) and $T_M$ (melting temperature) for aqueous $H_2SO_4$ solution. Chem Phys. Lett. 504, 34-36.

Vali, G. (1985). Nucleation terminology, J. Aerosol. Sci., 16(6), 575–576.

Vali, G. (2008). Repeatability and randomness in heterogeneous freezing nucleation, Atmos. Chem. Phys. Discuss., 8(1), 4059–4097.

Wang, H. and H. Gould. (2007). Homogeneous and heterogeneous nucleation in Lennard-Jones liquids. Phys. Rev. E 76, 031604.

Wang, G. M. and A. D. J Haymet. (1998). Trehalose and other sugar solutions at low temperature—modulated differential scanning calorimetry (Mdsc). J. Phys. Chem. B, 102, 5341–5347.

Wei, S., X. Xiaobin, Z. Hong and X. Chuanxiang. (2008). Effects of dipole polarization of water molecules on ice formation under and electrostatic field. Cryobiology 56, 93-99.

Wharton, D. A. (2011). Cold tolerance of New Zealand alpine insects. J. Insect Physiology 57, 1090-1095.

Wharton, D. A., Mutch J. S., Wilson, P. W., Marshall, C. J. and M. Lim. (2004). A simple ice nucleation spectrometer, CryoLetters 25, 335-340.

Wilson, P. W. and A. D. J. Haymet. (2009). Effect of solutes on the heterogeneous nucleation temperature of supercooled water: an experimental determination. Phys. Chem. Chem. Phys., 11, 2679 - 2682.

Wilson, P. W. J. W. Arthur and A. D. J Haymet. (1999). Ice premelting during differential scanning calorimetry. Biophys. J., 77, 2850–2855.

Wilson, P. W., Heneghan, A. F. and A. D. J. Haymet. (2003). Ice nucleation in Nature: supercooling point measurement and the role of heterogeneous nucleation. Cryobiology 46, 88-98.

Wilson, P. W. and J. P. Leader. (1995). Stabilization of supercooled fluids by thermal hysteresis proteins. Biophys. J. 68, 2098-2107.

Wilson, P. W., K. Osterday and A.D.J Haymet. (2009). The effects of electric field on ice nucleation are masked by the inherent stochastic nature of nucleation. CryoLetters 30, 2, 96-99.

Wilson, P.W., K. E. Osterday, A. F. Heneghan and A. D. J. Haymet. (2010). Effects of Type 1 antifreeze proteins on the heterogeneous nucleation of aqueous solutions. J. Biol. Chem. 285, 34741-34745.

Zachariassen, K. E. (1985). Physiology of cold tolerance in insects. Phys. Rev. 65, 799-832.

Zachariassen, K. E. and E. Kristiansen. (2000). Ice nucleation and antinucleation in Nature. Cryobiology 41, 257-279.

Zobrist, B., C. Marcolli, T. Peter and T. Koop. (2008). Heterogeneous ice nucleation in aqueous solutions: the role of water activity. J. Phys. Chem. A 112, 3965-3975.

# 2

# Supercooling and Freezing Tolerant Animals

David A. Wharton
*Department of Zoology, University of Otago,*
*Dunedin*
*New Zealand*

## 1. Introduction

Subzero temperatures may adversely affect animals by their direct lethal effects and by the damage caused by ice formation (Ramløv, 2000). Animals deal with the latter by three basic strategies. Freeze avoiding animals prevent ice formation in their bodies and supercool, keeping their body fluids liquid at temperatures below their melting point, but die if freezing occurs. In contrast, freezing tolerant animals can survive ice forming inside their bodies (Lee, 2010). Although both these categories of cold tolerance have been further subdivided (Bale, 1993; Sinclair, 1999) freeze avoidance and freezing tolerance are still recognised as fundamental cold tolerance strategies (Wharton, 2011a). In the third mechanism, cryoprotective dehydration which is found mainly in soil-dwelling animals, the body fluids remain unfrozen whilst surrounded by frozen soil. Since ice has a lower vapour pressure than liquid water at the same temperature the animal dehydrates and lowers the melting point of its body fluids, thus preventing freezing (Lee, 2010). In this review I examine the role of ice nucleation and supercooling in the main groups of freezing tolerant animals.

## 2. Freezing tolerant animals

Freezing tolerance has been most extensively studied in insects. It has been demonstrated in six insect orders, in which it appears to have evolved independently (Sinclair et al., 2003). Freeze tolerance is the dominant cold tolerance strategy in Southern Hemisphere insects, being found in 77% of cold hardy Southern Hemisphere insects (Sinclair & Chown, 2005). Amongst non-insect arthropods, however, freeze avoidance is the dominant cold tolerance strategy and freezing tolerance has only been demonstrated in a single species of centipede (Tursman et al., 1994), in an aquatic subterranean crustacean (Issartel et al., 2006) and in intertidal barnacles (Storey & Storey, 1988).

In nematodes, *Panagrolaimus davidi* from Antarctica is freezing tolerant and has survived temperatures down to -80°C (Wharton, 2011a). Other species of nematodes seem to have more modest cold tolerance abilities (Smith et al., 2008), although some have a small amount of freezing tolerance (Hayashi & Wharton, 2011). Tardigrades are thought to be freezing tolerant (Hengherr et al., 2009), although the role played by inoculative freezing and whether ice formation is intracellular or extracellular is yet to be determined in this phylum.

Some rotifers can survive at very low temperatures (Newsham et al., 2006) but their cold tolerance mechanisms have not been studied.

Some intertidal molluscs have an ability to tolerate freezing, at least in a part of their tissues. This depends upon inoculative freezing from the surrounding seawater, the seasonal production of INAs and the proportion of tissue that is frozen (Ansart & Vernon, 2003). Most earthworms avoid freezing by migrating to deeper soil layers during winter and thus avoiding contact with frozen soil. However, some species permanently inhabit the surface layers of the soil and if frost occurs they must survive contact with ice. A few earthworm species are freezing tolerant and, although some can remain unfrozen at -1°C, appear to rely on inoculative freezing from the surrounding soil to ensure freezing at a high subzero temperature (Holmstrup, 2003; Holmstrup & Overgaard, 2007).

Amongst vertebrates, several species of North American and European frogs are freezing tolerant and this cold tolerance strategy appears to have evolved several times amongst anurans (Voituron et al., 2009). Freezing tolerance has also been reported in the Siberian salamander (Berman et al., 1984) and in a single species of frog from the Southern Hemisphere (Bazin et al., 2007). In reptiles, hatchling turtles that overwinter in terrestrial hibernacula (the nest in which they were born) have the ability to tolerate freezing but the role this plays under natural conditions has been the subject of debate (Costanzo et al., 2008;

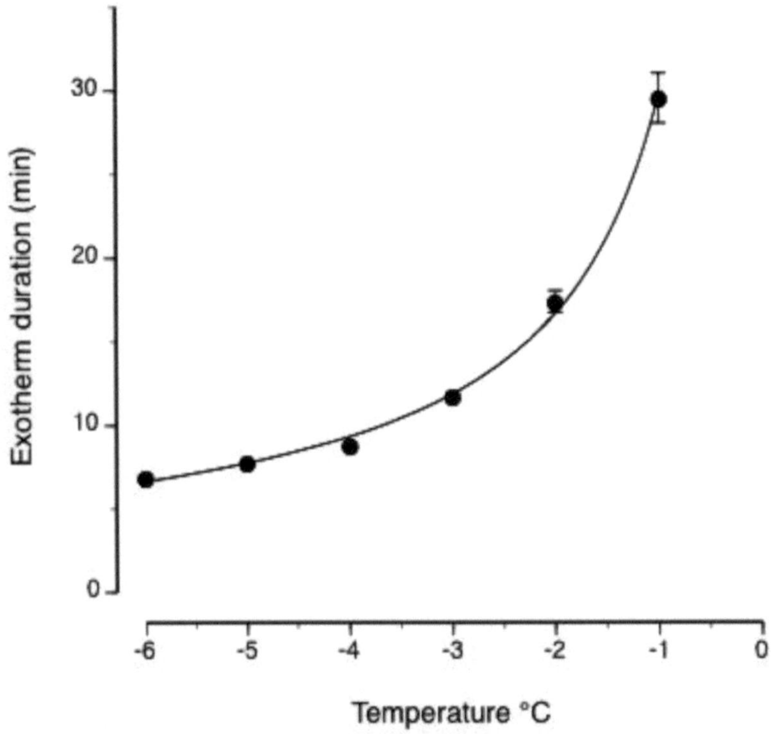


Fig. 1. The effect of temperature on the exotherm duration of 10µl suspensions of the Antarctic nematode *Panagrolaimus davidi*. $N = 4$, bars are standard errors. The line is the fit to the equation $f(x) = \exp(3.39) * (x^{-0.83})$, $R^2 = 0.996$. Redrawn from (Wharton et al., 2002).

Packard & Packard, 2004). Some species of lizards and snakes can survive partial freezing of their bodies but die once a critical level of ice is exceeded (Storey, 2006). The European common lizard, *Lacerta vivipara*, can survive 50% of its body water freezing for at least 24 h, which is an ecologically-relevant level of freezing tolerance (Voituron et al., 2002). No fish, mammal or bird has been reported to survive more than a small amount of their body fluids freezing.

## 3. Most freezing tolerant animals limit supercooling

In contrast to freeze avoiding animals (see Ramløv, this volume), most freezing tolerant animals prevent extensive supercooling and encourage ice formation at a high subzero temperature. A clue to the reason why they do this can be seen from the relationship between the rate of ice formation and temperature. Figure 1 shows that the rate at which water freezes (or in this case a suspension of nematodes in water; with the duration of the exotherm produced during the freezing of the sample used as a measure of the freezing rate) decreases dramatically as the temperature approaches its melting point. Figure 2 compares exotherms in 50µl of nematode suspension where freezing was initiated at -1.3°C and -11.0°C. The exotherm duration is about 4.5 times longer when freezing is initiated at -1.3°C than at -11°C. Perhaps of more significance is that at -1.3°C the temperature becomes elevated to the melting point of the suspension (-0.3°C) and remains there until the freezing process is completed, indicating that the spread of ice through the sample is slow. At -11°C the temperature fails to reach the melting point of the suspension and declines rapidly from the maximum temperature reached (-0.9°C), indicating the rapid spread of ice through the sample.

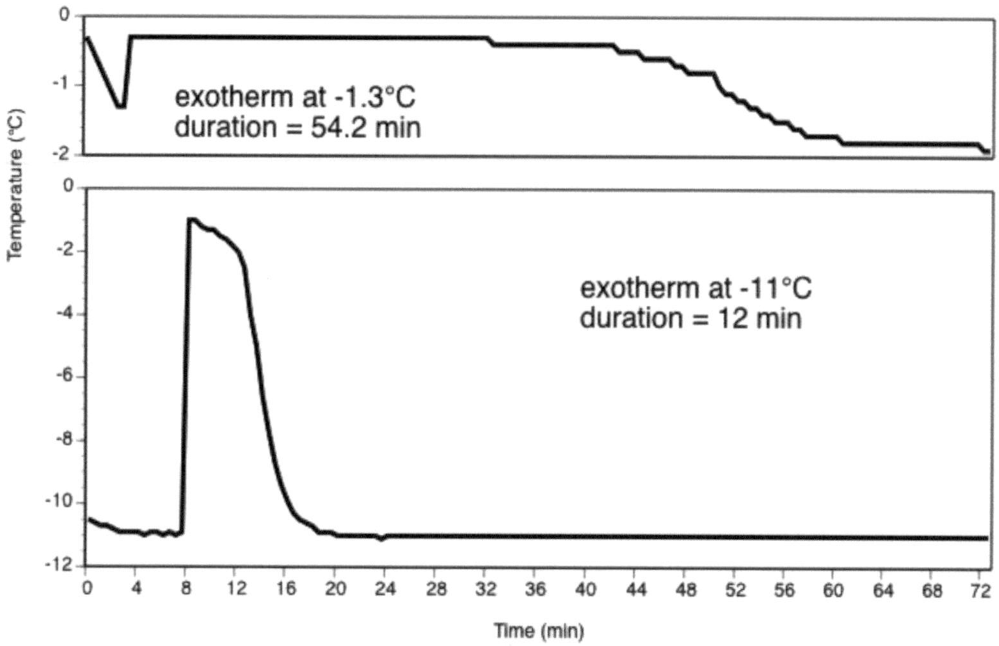


Fig. 2. Exotherms from 50µl suspensions of the Antarctic nematode *Panagrolaimus davidi*, with freezing initiated at -1.3°C (top) and -11.0°C (bottom) (data from Wharton et al., 2002).

In the moderately freezing tolerant insect *Celatoblatta quinquemaculata* cooled to different temperatures (Fig. 3), exotherms are similar since freezing is initiated at a relatively high subzero temperature by ice nucleators in the gut or haemolymph of the cockroach (Worland et al., 1997). The temperature at which the insect freezes spontaneously (the whole body supercooling point, SCP or temperature of crystallization, $T_c$) is not significantly different between animals frozen to -5, -8 or -12°C, with a mean SCP of -4.0±0.2°C (mean±se, *N* = 12) and the temperature becomes elevated to the melting point of the animals' body fluids in each case (-0.5 to -1.4°C) (Worland et al., 2004).

Across a range of freezing tolerant animals freezing tends to occur at high subzero temperature, the rate of ice formation is slow and it takes a long time for the exotherm associated with the freezing event to be completed (Table 1). Perhaps not surprisingly these parameters are broadly correlated with the size of the animal.

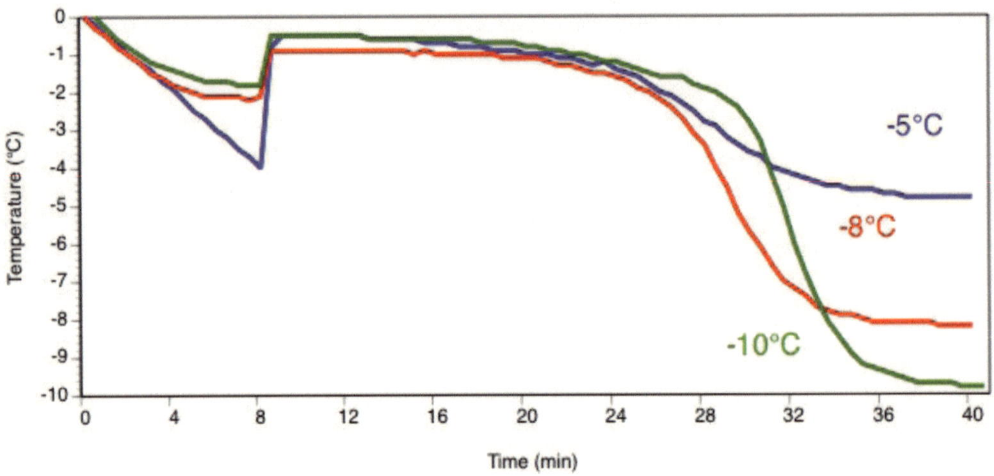


Fig. 3. Exotherms from the New Zealand alpine cockroach cooled to -5, -8 or -10°C at 0.5°C $min^{-1}$ (data from Worland et al., 2004).

## 4. Freezing tolerance types in insects

Freezing tolerant insects vary in the relationship between the SCP and their lower lethal temperature (LLT). In order to be categorized as freezing tolerant the LLT of the insect must be lower than its SCP, and the insect survives ice formation in its body proceeding to completeness. In some species the LLT is only a few degrees below the SCP, which occurs at a high subzero temperature. These have been called 'moderately freezing tolerant' insects (Sinclair, 1999), although it would be more correct to call them freezing tolerant but moderately cold tolerant. In other freezing tolerant insects the LLT is many degrees below the SCP, which occurs at a high subzero temperature. These have been called (Sinclair, 1999) 'strongly freezing tolerant' (freezing tolerant and strongly cold tolerant). A final type has a low SCP but the LLT is a few degrees lower still: 'freezing tolerant with a low SCP' (Sinclair, 1999). 'Partial freezing tolerance' has also been proposed as a category, for those insects that will survive some ice forming in their bodies but die if the exotherm is completed (Sinclair, 1999). These insects are not freezing tolerant (since they do not survive completion of the freezing process) but they could represent a stage in the evolution of freezing tolerance (Hawes & Wharton, 2010; Sinclair, 1999).

| Group | Animal | mass | $T_{min}$ °C | $T_c$ °C | Exotherm duration | Reference |
|---|---|---|---|---|---|---|
| Vertebrates | *Rana sylvatica* | ~14g | -2 | -2.2 ±0.1 | 14 to 20+ h | Layne & Lee, 1987 |
| | *Litoria ewingii* | ~1g | -2 | -1.7 ±0.3[1] | 1.7 ±0.3 h | Bazin et al., 2007 |
| | *Chrysemys picta* | 3-6g | -2.5 | -2.5 ±0.46 | ~12 h[2] | Churchill & Storey, 1992 |
| Insects | *Celatoblatta quinquemaculata* | ~0.1g | -5 | -4.6 ±0.1 | 15.1 ±1.2 min | Worland et al., 2004 |
| | *Hemideina maori* | ~5g | -5 | -0.8 to -2.5 | 10 h[2] | Ramløv & Westh, 1993 |
| | *Eurosta solidaginis* | ~0.05g | -23 | -10.6 | 48 h[2] | Lee & Lewis, 1985 |
| Nematode | *Panagrolaimus davidi*[3] | <1μg | $T_c$ | -26.5 ±0.9 | 13.0 ±2.0 s | Wharton & Block, 1997 |

[1]cooled on a dry substrate at 1°C $h^{-1}$ until freezing occurred
[2]time to maximum ice formation, [3]for single nematodes free of surface water in liquid paraffin
$T_{min}$ minimum temperature, $T_c$ temperature of crystallization.

Table 1. The freezing characteristics of some freezing tolerant animals

Moderately freezing tolerant insects predominate in the Southern Hemisphere, where they are associated with climates that have mild winters but where temperatures are unpredictable and can fall below 0°C at any time of the year (Chown & Sinclair, 2010; Sinclair, et al., 2003). These insects have relatively high SCPs and a large proportion of their body water is converted into ice. In *C. quinquemaculata* 74% of their water freezes (Block et al., 1998) and in *H. maori* 82% freezes (Ramløv & Westh, 1993). However, in both of these insects their LLT is only about 6°C below their SCP (Wharton, 2011b). There must be further lethal events at lower temperatures after ice formation has been completed. Given the high proportion of ice that is formed initially it seems unlikely that this is due to further compartments freezing; as has been observed in an Alaskan fungus gnat (Sformo et al., 2009).

Using a live/dead cell stain Worland et al. (Worland, et al., 2004) showed that a high proportion of cells in the midgut, fat body and Malpighian tubules of *C. quinquemaculata* survived freezing at -5°C. At lower temperatures (-8°C, -12°C) the proportion of dead cells increases, with the fat body cells being the most sensitive tissue of those tested. In *H. maori* the Malpighian tubule cells are more sensitive to low temperatures than are the fat body cells but the survival of both declines with temperature (Sinclair & Wharton, 1997). These results suggest that the LLT of these insects reflects the accumulation of damage in a critical tissue. In the strongly freezing tolerant insect *E. solidaginis* tissues vary in the ability of the cells to survive freezing, with the cells of the alimentary system being the most resistant (Yi & Lee, 2003). Again it is the most sensitive tissue that will set the lower limit on the survival of the organism.

## 5. Ice nucleation in freezing tolerant animals

In contrast to freeze avoiding animals, that eliminate or mask sources of ice nucleation, freezing tolerant animals allow and encourage ice nucleation. Some freezing tolerant insects produce proteins or lipoproteins that have ice nucleating activity. These ensure that freezing occurs at a relatively high subzero temperature. They may also control the site of ice formation so that it occurs in the haemocoel, preventing potentially fatal intracellular freezing (Duman et al., 2010). The freezing tolerant Southern Hemisphere frog *Litoria ewingi* has ice nucleators in its skin secretions (Rexer-Huber et al., 2011), which ensure that this winter-active and largely terrestrial frog will freeze at a very high subzero temperature (-1.7°C) even on a dry substrate (Bazin, et al., 2007).

Moderately freezing tolerant insects continue to feed during the winter, ensuring the year-round presence of food and microorganisms in their gut that could act as ice nucleators. *Celatoblatta quinquemaculata* and *H. maori* have ice nucleators in their haemolymph, gut contents and faeces (Wilson & Ramløv, 1995; Worland, et al., 1997). The nucleating activity of the faeces is greater than that of the haemolymph (Sinclair et al., 1999; Worland, et al., 1997). This suggests that the gut is the primary site of ice nucleation, with nucleators in the haemolymph providing a back-up system if the gut is empty (Worland, et al., 1997).

The strongly freezing tolerant insect *E. solidaginis* forms a non-feeding dormant larval stage overwinter, surviving within the gall it induces in the stem of its host plant (Baust & Nishino, 1991). When the water content of the gall is high the larvae freeze by ice inoculation from the surrounding plant tissue (Lee & Hankison, 2003). As the autumn and winter progresses the galls dry out, inoculative freezing decreases and the insects rely on endogenous nucleators. These include calcium phosphate spherules that accumulate in the Malpighian tubules of overwintering larvae. These spherules, and the insect's fat body cells, have ice nucleating activity that ensure that the larvae freeze at -8°C to -10°C (Mugnano et al., 1996).

Although some nematodes can survive desiccation, for growth and reproduction to occur at least a film of water must be present. Nematodes, and animals that live in similar habitats (such as tardigrades and rotifers), are likely to be faced with the risk of inoculative freezing from ice in their surroundings. Few species have been examined in this respect, but in those that have (*Panagrolaimus davidi* and *Panagrellus redivivus*) they have little ability to resist inoculative freezing (Hayashi & Wharton, 2011; Wharton & Ferns, 1995; Wharton et al., 2003). This also is the case in the infective larvae of the insect parasitic nematode *Steinernema feltiae* (Farman & Wharton, unpublished results) and the free-living Antarctic nematode *Plectus murrayi* (Raymond, 2010). In *P. davidi* inoculative freezing occurs via body openings, especially the excretory pore (Wharton & Ferns, 1995) and endogenous ice nucleators are absent (Wharton & Worland, 1998). However, if freezing of the media occurs at a high subzero temperature (-1°C) inoculative freezing does not occur in *P. davidi* and the nematode can survive by cryoprotective dehydration (Wharton, et al., 2003). In *P. redivivus*, however, inoculative freezing occurs in some individuals even at -1°C and the small amount of cold tolerance that this species possesses is largely due to freezing tolerance, although those few nematodes that remain unfrozen survive by cryoprotective dehydration (Hayashi & Wharton, 2011).

Other freezing tolerant animals can use cryoprotective dehydration as an alternative strategy to freezing tolerance, especially under conditions where the chances of inoculative freezing is reduced (such as in soil of low water content). This has been reported in freezing-tolerant earthworms (Pedersen & Holmstrup, 2003) and in the Antarctic midge, *Belgica antarctica* (Elnitsky et al., 2008).

Some freezing tolerant arthropods appear to rely on inoculative freezing for survival. The centipede *Lithobius forficatus* freezes at a temperature just below the melting point of their haemolymph (about -1°C) by inoculative freezing when in contact with ice and survives, but if it supercools to -7°C and below it dies when it freezes (Tursman, et al., 1994). Caterpillars of the moth *Cisseps fulvicolis* also require inoculative freezing at a high subzero temperature to tolerate freezing (Fields & McNeil, 1986). Diapausing larvae of the fly *Chymomyza costata* can survive to low temperatures better if freezing is initiated by inoculative freezing at -2°C, than if they are allowed to supercool (Shimada & Riihimaa, 1988).

Inoculative freezing may also be an important factor in the cold tolerance mechanisms of vertebrate ectotherms. The skin of frogs has a high permeability to water and if they are cooled in contact with a moist natural substrate (such as soil or leaf mould) they freeze by inoculative freezing when ice forms in the substrate. Soil contains abundant ice nucleators that initiate freezing at a high subzero temperature (Costanzo et al., 1999). The skin of hatchling turtles is much less permeable to water than that of frogs but inoculative freezing can still occur via body openings, such as the eyes, ears, nose, cloaca, umbilicus, mouth and anus. Given the high levels of ice nucleators in their natural substrate this necessitates a level of freezing tolerance (Costanzo, et al., 2008) and inoculative freezing may be required to ensure survival over winter (Baker et al., 2006).

## 6. Intracellular freezing

The use of ice nucleators to induce freezing at high subzero temperature in the body fluids of many freezing tolerant animals is thought to ensure that freezing occurs extracellularly (Duman, et al., 2010). Intracellular freezing is thought to be fatal due to the mechanical disruption of cells by the expansion of water as it freezes, the puncturing of membranes by ice crystals or the redistribution of ice crystals (recrystallization) after freezing and during thawing (Acker & McGann, 2001; Muldrew et al., 2004). However, some examples of survival of intracellular freezing have been discovered in particular cells and tissues of some freezing tolerant animals (Sinclair & Renault, 2010). The only animal shown to survive extensive intracellular freezing throughout its body is the Antarctic nematode *P. davidi* (Wharton & Ferns, 1995). Some other nematodes have now been shown to have at least some ability to survive intracellular freezing, including *Steinernema feltiae* (Farman & Wharton, unpublished results) and *Plectus murrayi* (Raymond, 2010).

## 7. Conclusions

An animal is said to be freezing tolerant if it can survive the freezing and thawing of a biologically significant amount of its body water under thermal and temporal conditions that reflect its exposure to low temperatures in nature (Baust, 1991). In the laboratory this is demonstrated by the ability to survive ice formation, or the exotherm associated with freezing, going to completion but is harder to demonstrate in a natural situation (Costanzo, et al., 2008). Animals from many different phyla are freezing tolerant, including both invertebrate and vertebrate ectotherms. Most freezing tolerant animals, but not all, freeze at a high subzero temperature; with the SCP being controlled by the production of ice nucleators, the retention of food and bacteria in the gut or by allowing inoculative freezing. This ensures that the freezing process is slow and gentle, allowing the animal to adjust to the changing physiological conditions. Ice formation is usually thought to be extracellular but this has rarely been examined and survival of intracellular freezing may be much more widespread amongst freezing tolerant animals than we currently realise.

## 8. References

Acker, J.P. & McGann, L.E. (2001) Membrane damage occurs during the formation of intracellular ice. *CryoLetters*, Vol. 22, 241-254.

Ansart, A. & Vernon, P. (2003) Cold hardiness in molluscs. *Acta Oecologica*, Vol. 24, 95-102.

Baker, P.J., Costanzo, J.P., Herlands, R., Wood, R.C. & Lee, R.E. (2006) Inoculative freezing promotes winter survival in hatchling diamondback terrapin, *Malaclemys terrapin*. *Canadian Journal of Zoology*, Vol. 84, 116-124.

Bale, J.S. (1993) Classes of insect cold hardiness. *Functional Ecology*, Vol. 7, 751-753.

Baust, J.G. (1991) The freeze tolerance oxymoron. *CryoLetters*, Vol. 12, 1–2.

Baust, J.G. & Nishino, M. (1991) Freezing tolerance in the goldenrod gall fly (*Eurosta solidaginis*). In: *Insects at Low Temperatures*, Lee, R.E. & Denlinger, D.L. (Eds.), pp. 260-275, Chapman and Hall, New York and London.

Bazin, Y., Wharton, D.A. & Bishop, P.J. (2007) Cold tolerance and overwintering of an introduced New Zealand frog, the brown tree frog (*Litoria ewingii*). *CryoLetters*, Vol. 28, 347-358, ISSN 0143-2044.

Berman, D.I., Leirikh, A.N. & Mikhailova, E.I. (1984) Winter hibernation of the Siberian salamander *Hynobius keyserlingi*. *Journal of Evolutionary Biochemistry and Physiology*, Vol. 20, 323–327.

Block, W., Wharton, D.A. & Sinclair, B.J. (1998) Cold tolerance of a New Zealand alpine cockroach, *Celatoblatta quinquemaculata* (Blattoidea: Blattidae). *Physiological Entomology*, Vol. 23, 1-6.

Chown, S.L. & Sinclair, B.J. (2010) The macrophysiology of insect cold hardiness. In: *Low Temperature Biology of Insects*, Denlinger, D.L. & Lee, R.E. (Eds.), pp. 191-222, Cambridge University Press, ISBN 978-0-521-88635-2, Cambridge.

Churchill, T.A. & Storey, K.B. (1992) Natural freezing survival by painted turtles *Chrysemys picta marginata* and *C. picta belli*. *American Journal of Physiology*, Vol. 262, R530-R537.

Costanzo, J.P., Bayuk, J.M. & Lee, R.E. (1999) Inoculative freezing by environmental ice nuclei in the freeze-tolerant wood frog, *Rana sylvatica*. *Journal of Experimental Zoology*, Vol. 284, 7-14.

Costanzo, J.P., Lee, R.E. & Ultsch, G.R. (2008) Physiological ecology of overwintering in hatchling turtles. *Journal of Experimental Zoology*, Vol. 309A, 297-379, ISSN 1932-5223.

Duman, J.G., Walters, K.R., Sformo, T., Carrasco, M.A., Nickell, P.A., Lin, X. & Barnes, B.M. (2010) Antifreeze and ice-nucleator proteins. In: *Low Temperature Biology of Insects*, Denlinger, D.L. & Lee, R.E. (Eds.), pp. 59-90, Cambridge University Press, ISBN 978-0-521-88635-2, Cambridge.

Elnitsky, M.A., Hayward, S.A.L., Rinehart, J.P., Denlinger, D.L. & Lee, R.E. (2008) Cryoprotective dehydration and the resistance to inoculative freezing in the Antarctic midge, *Belgica antarctica*. *Journal of Experimental Biology*, Vol. 211, 524-530, ISSN 0022-0949.

Fields, P.G. & McNeil, J.N. (1986) Possible dual cold-hardiness strategies in *Cisseps fulvicolis* (Lepidoptera: Arctiidae). *Canadian Entomologist*, Vol. 118, 1309-1311.

Hawes, T.C. & Wharton, D.A. (2010) Tolerance of freezing in caterpillars of the New Zealand Magpie moth (*Nyctemera annulata*). *Physiological Entomology*, Vol. 35, 296-300, ISSN 0307-6962.

Hayashi, M. & Wharton, D.A. (2011) The oatmeal nematode *Panagrellus redivivus* survives moderately low temperatures by freezing tolerance and cryoprotective dehydration. *Journal of Comparative Physiology B*, Vol. 181, 335-342, ISSN 0174-1578.

Hengherr, S., Worland, M.R., Reuner, A., Brummer, F. & Schill, R.O. (2009) Freeze tolerance, supercooling points and ice formation: comparative studies on the subzero temperature survival of limno-terrestrial tardigrades. *Journal of Experimental Biology*, Vol. 212, 802-807, ISSN 0022-0949.

Holmstrup, M. (2003) Overwintering adaptations in earthworms. *Pedobiologia*, Vol. 47, 504-510.

Holmstrup, M. & Overgaard, J. (2007) Freeze tolerance in *Aporrectodea caliginosa* and other earthworms from Finland. *Cryobiology*, Vol. 55, 80-86.

Issartel, J., Voituron, Y., Odagescu, V., Baudot, A., Guillot, G., Ruaud, J.P., Renault, D., Vernon, P. & Hervant, F. (2006) Freezing or supercooling: how does an aquatic subterranean crustacean survive exposures at subzero temperatures? *Journal of Experimental Biology*, Vol. 209, 3469-3475.

Layne, J.R. & Lee, R.E. (1987) Freeze tolerance and the dynamics of ice formation in wood frogs (*Rana sylvatica*) from southern Ohio. *Canadian Journal of Zoology*, Vol. 65, 2062-2065.

Lee, R.E. & Lewis, E.A. (1985) Effect of temperature and duration of exposure on tissue ice formation in the gall fly, *Eurosta solidaginis* (Diptera, Tephritidae). *CryoLetters*, Vol. 6, 25-34.

Lee, R.E. & Hankison, S.J. (2003) Acquisition of freezing tolerance in early autumn and seasonal changes in gall water content influence inoculative freezing of gall fly larvae, *Eurosta solidaginis* (Diptera, Tephritidae). *Journal of Insect Physiology*, Vol. 49, 385-393.

Lee, R.E. (2010) A primer on insect cold tolerance. In: *Low Temperature Biology of Insects*, Denlinger, D.L. & Lee, R.E. (Eds.), pp. 3-34, Cambridge University Press, ISBN 978-0-521-88635-2, Cambridge.

Mugnano, J.A., Lee, R.E. & Taylor, R.T. (1996) Fat body cells and calcium phosphate spherules induce ice nucleation in the freeze-tolerant larvae of the gall fly *Eurosta solidaginis* (Diptera, Tephritidae). *Journal of Experimental Biology*, Vol. 199, 465-471.

Muldrew, K., Acker, J.P., Elliott, J.A.W. & McGann, L.E. (2004) The water to ice transition: Implications for living cells. In: *Life in the Frozen State*, Fuller, B.J., Lane, N. & Benson, E.E. (Eds.), pp. 67-108, CRC Press Inc, ISBN 0415247004, Boca Raton.

Newsham, K.K., Maslen, N.R. & McInnes, S.J. (2006) Survival of antarctic soil metazoans at -80 degrees C for six years. *CryoLetters*, Vol. 27, 291-294.

Packard, G.C. & Packard, M.J. (2004) To freeze or not to freeze: adaptations for overwintering by hatchlings of the North American painted turtle. *Journal of Experimental Biology*, Vol. 207, 2897-2906.

Pedersen, P.G. & Holmstrup, M. (2003) Freeze or dehydrate: Only two options for the survival of subzero temperatures in the arctic enchytraeid *Fridericia ratzeli*. *Journal of Comparative Physiology B*, Vol. 173, 601-609.

Ramløv, H. & Westh, P. (1993) Ice formation in the freeze tolerant alpine weta *Hemideina maori* Hutton (Orthoptera; Stenopelmatidae). *CryoLetters*, Vol. 14, 169-176.

Ramløv, H. (2000) Aspects of natural cold tolerance in ectothermic animals. *Human Reproduction*, Vol. 15, 26-46.

Raymond, M.R. (2010) *Cold-temperature Adaptation in Nematodes from the Victoria Land Coast, Antarctica.* PhD thesis, University of Otago, New Zealand, 312pp.

Rexer-Huber, K.M.J., Bishop, P.J. & Wharton, D.A. (2011) Skin ice nucleators and glycerol in the freezing-tolerant frog *Litoria ewingii*. *Journal of Comparative Physiology B*, Vol. 181, 781-792, ISSN 0174-1578.

Sformo, T., Kohl, F., McIntyre, J., Kerr, P., Duman, J.G. & Barnes, B.M. (2009) Simultaneous freeze tolerance and avoidance in individual fungus gnats, *Exechia nugatoria*. *Journal of Comparative Physiology B*, Vol. 179, 897-902, ISSN 0174-1578.

Shimada, K. & Riihimaa, A. (1988) Cold acclimation, inoculative freezing and slow cooling: essential factors contributing to the freezing-tolerance in diapausing larvae of *Chymomyza costata* (Diptera: Drosophilidae). *CryoLetters*, Vol. 9, 5-10.

Sinclair, B.J. & Wharton, D.A. (1997) Avoidance of intracellular freezing by the freezing-tolerant New Zealand alpine weta *Hemideina maori* (Orthoptera: Stenopelmatidae). *Journal of Insect Physiology*, Vol. 43, 621-625.

Sinclair, B.J. (1999) Insect cold tolerance: How many kinds of frozen? *European Journal of Entomology*, Vol. 96, 157-164.

Sinclair, B.J., Worland, M.R. & Wharton, D.A. (1999) Ice nucleation and freezing tolerance in New Zealand alpine and lowland weta, *Hemideina* spp. (Orthoptera; Stenopelmatidae). *Physiological Entomology*, Vol. 24, 56-63.

Sinclair, B.J., Addo-Bediako, A. & Chown, S.L. (2003) Climatic variability and the evolution of insect freeze tolerance. *Biological Reviews*, Vol. 78, 181-195.

Sinclair, B.J. & Chown, S.L. (2005) Climatic variability and hemispheric differences in insect cold tolerance: support from southern Africa. *Functional Ecology*, Vol. 19, 214-221.

Sinclair, B.J. & Renault, D. (2010) Intracellular ice formation in insects: Unresolved after 50 years? *Comparative Biochemistry and Physiology A*, Vol. 155, 14-18, ISSN 1095-6433.

Smith, T., Wharton, D.A. & Marshall, C.J. (2008) Cold tolerance of an Antarctic nematode that survives intracellular freezing: comparisons with other nematode species. *Journal of Comparative Physiology B*, Vol. 178, 93-100, ISSN 0174-1578.

Storey, K.B. & Storey, J.M. (1988) Freeze tolerance in animals. *Physiological Reviews*, Vol. 68, 27-84.

Storey, K.B. (2006) Reptile freeze tolerance: metabolism and gene expression. *Cryobiology*, Vol. 52, 1-16.

Tursman, D., Duman, J.G. & Knight, C.A. (1994) Freeze tolerance adaptations in the centipede, *Lithobius fortificatus*. *Journal of Experimental Zoology*, Vol. 268, 347-353.

Voituron, Y., Storey, J.M., Grenot, C. & Storey, K.B. (2002) Freezing survival, body ice content and blood composition of the freeze-tolerant European common lizard, *Lacerta vivipara*. *Journal of Comparative Physiology B*, Vol. 172, 71-76.

Voituron, Y., Barre, H., Ramlov, H. & Douady, C.J. (2009) Freeze tolerance evolution among anurans: Frequency and timing of appearance. *Cryobiology*, Vol. 58, 241-247, ISSN 0011-2240.

Wharton, D.A. & Ferns, D.J. (1995) Survival of intracellular freezing by the Antarctic nematode *Panagrolaimus davidi*. *Journal of Experimental Biology*, Vol. 198, 1381-1387.

Wharton, D.A. & Block, W. (1997) Differential scanning calorimetry studies on an Antarctic nematode (*Panagrolaimus davidi*) which survives intracellular freezing. *Cryobiology*, Vol. 34, 114-121.

Wharton, D.A. & Worland, M.R. (1998) Ice nucleation activity in the freezing tolerant Antarctic nematode *Panagrolaimus davidi*. *Cryobiology*, Vol. 36, 279-286.

Wharton, D.A., Goodall, G. & Marshall, C.J. (2002) Freezing rate affects the survival of a short-term freezing stress in *Panagrolaimus davidi*, an Antarctic nematode that survives intracellular freezing. *CryoLetters*, Vol. 23, 5-10.

Wharton, D.A., Goodall, G. & Marshall, C.J. (2003) Freezing survival and cryoprotective dehydration as cold tolerance mechanisms in the Antarctic nematode *Panagrolaimus davidi*. *Journal of Experimental Biology*, Vol. 206, 215-221.

Wharton, D.A. (2011a) Cold tolerance. In: *Molecular and Physiological Basis of Nematode Survival*, Perry, R.N. & Wharton, D.A. (Eds.), pp. 182-204, CABI Publishing, ISBN 9781845936877, Wallingford.

Wharton, D.A. (2011b) Cold tolerance of New Zealand alpine insects. *Journal of Insect Physiology*, Vol. 57, 1090-1095, ISSN 0022-1910.

Wilson, P.W. & Ramløv, H. (1995) Hemolymph ice nucleating proteins from the New Zealand alpine weta *Hemideina maori* (Orthoptera: Stenopelmatidae). *Comparative Biochemistry and Physiology B*, Vol. 112, 535-542.

Worland, M.R., Sinclair, B.J. & Wharton, D.A. (1997) Ice nucleation activity in a New Zealand alpine cockroach *Celatoblatta quinquemaculata* (Dictyoptera: Blattidae). *CryoLetters*, Vol. 18, 327-334.

Worland, M.R., Wharton, D.A. & Byars, S.G. (2004) Intracellular freezing and survival in the freeze tolerant alpine cockroach *Celatoblatta quinquemaculata*. *Journal of Insect Physiology*, Vol. 50, 225-232.

Yi, S.X. & Lee, R.E. (2003) Detecting freeze injury and seasonal cold-hardening of cells and tissues in the gall fly larvae, *Eurosta solidaginis* (Diptera : Tephritidae) using fluorescent vital dyes. *Journal of Insect Physiology*, Vol. 49, 999-1004.

# 3

# Glass Transition Behavior of Aqueous Solution of Sugar-Based Surfactants

Shigesaburo Ogawa[1] and Shuichi Osanai[2]
[1]*Kyushu University*
[2]*Kanagawa University*
*Japan*

## 1. Introduction

Since the end of previous century, the role of petroleum as a raw material of synthetic surfactant gradually deflated due to the reasons such as decreasing of the relative abundance of petroleum, leading to soared prices of petroleum and increasing of carbon dioxide emission by heavy utilization of petroleum. Instead, the industries concerning in the surfactants and detergents are focusing on the utilization of biobased feedstocks, intermediates and products. Under these circumstances, the biobased surfactants derived from carbohydrate or sugar are highlighted.

Sugar-based surfactants commonly used for household products are frequently applied in foods, cosmetics and pharmaceutical industrial region (Rybinski, von W. & Hill . K. (1998). Hill, K. & Rhode O. (1999). Drummond, C. J.; Fong, C.; Krodkiewska, I.; Boyd, B. J. & Baker, I. J. A. (2003). Hill, K. & LeHen-Ferrenbach, C. (2007).). They are less toxic, highly biodegradable, and able to be readily formulated with other components. And it is well known that their representative nature is that they have ability to aggregate in an aqueous solution as well as conventional surfactants (Warr, G. G.; Drummond, C. J.; Grieser, F.; Ninham, B. W. & Evans, D. F. (1986). Auvray, X.; Petipas, C. & Anthore, R. (1995). Söderberg, I.; Drummond, C. J.; Furlong, D. N.; Godkin, S. & Matthews, B. (1995). Hoffmann, B. & Platz, G. (2001). Kocherbitov, V. & Söderman, O. (2003). Imura, T.; Hikosaka, Y.; Worakitkanchanakul, W.; Sakai, H.; Abe, M.; Konishi, M.; Minamikawa, H. & Kitamoto, D. (2007). Hato, M.; Minamikawa, H. & Kato T. (2007).). The morphology of the aggregate extends over ranges from the isotropic micelle solution to the liquid crystal such as hexagonal, cubic, lamella and sponge phases.

Numerous phase diagrams of the amphiphiles, which describe the aggregative behavior of the compound, are exhibited in terms of concentration and temperature. We are able to see those of the anionic, cationic and nonionic surfactant, but the diagram under 0 °C especially in the frozen state was not reported so much. Among such studies, cationic surfactant, octyl trimethylammonium bromide is reported to be able to lower the freezing point of ice effectively due to the presence of their ionic head group (Fukada, K.; Matsuzaka, Y.; Fujii, M.; Kato, T. & Seimiya, T. (1998).). Similarly, nonionic surfactant such as polyoxyethylene glycol decyl ($C_{10}E_m$ ; m = 4-8) and dodecyl ($C_{12}E_m$ ; m = 5, 6 and 8) ether were reported to crystallize ice below −11 °C or −4.5 °C, respectively (Andersson, B. & Olofsson, G. (1987). Nibu, Y.; Suemori, T. & Inoue T. (1997). Nibu, Y. & Inoue, T. (1998a, 1998b). Zheng, L. Q.; Suzuki, M. & Inoue, T. (2002). Zheng, L.; Suzuki, M.; Inoue, T. & Lindman, B. (2002).). Contrary to this, the phase diagram of sugar-based surfactant seems to be uncompleted particularly under supercooled conditions under 0 °C.

Some nonionic surfactants were not used as a curative agent but a plasticizer because they showed the glass transition temperature ($T_g$) at low temperature region (Jensen, R. E.; O'Brien, E.; Wang, J.; Bryant, J.; Ward, T. C.; James, L. T. & Lewis, D. A. (1998); Amim, J.; Kawano, Y. & Petri, D. F. S. (2009).). Tween 40, poly(oxyethylene) sorbitan monopalmitate which have 20 EO units in the molecule was reported to possess $T_g$ at −61 °C. Triton X-100 showed its $T_g$ at −59 °C. Ethylene oxide surfactant such as hexahydrofarnesyl ethylene oxide surfactants (EO = 1 - 8) exhibited their $T_g$ at low temperatures below −80 °C (Fong, C.; Weerawardena, A.; Sagnella, S. M.; Mulet, X.; Krodkiewska, I.; Chong, J. & Drummond, C. J. (2011).). In addition to this, there is a report that says sugar derivatives containing a hydrophobic group are applicable as a plasticizer. Gill stated that when such a sugar derivative was added to the corresponding free sugar, $T_g$ of the mixture tended to lower than those of the free sugar system (Gill, I. & Valivety, R. (2000a, 2000b).). Here, the sugar given hydrophobicity worked as a plasticizer for a free sugar. On the other hand, when the other component which possessed much lower $T_g$ than that of the sugar derivative was mixed in the system, the existing sugar derivative did not necessarily work as a plasticizer.

Although it had been scarcely studied about the glass-forming property of sugar-based surfactants, but nowadays, much attention is being denoted to their interesting characteristics. It has been reported that *n*-alkyl glycosides such as α-D-glucosides, β-D-maltosides, β-D-maltotrioside and sucrose fatty acid esters formed a glass state under anhydrous conditions (Hoffmann, B.; Milius, W.; Voss, G.; Wunschel, M.; van Smaalen, S.; Diele, S. & Platz G. (2000). Kocherbitov, V. & Söderman, O. (2004). Ericsson, C. A.; Ericsson, L. C.; Kocherbitov, V.; Söderman, O. & Ulvenlund, S. (2005). Ericsson, C. A.; Ericsson, L. C. & Ulvenlund, S. (2005). Szűts, A.; Pallagi, E.; Regdon, G. Jr; Aigner, Z.; Szabó-Révész, P. (2007).). Their $T_g$ increased from −12.4 °C of *n*-heptyl α-D-glucopyranoside to 100 °C of *n*-dodecyl β-D-maltotrioside in proportional to the number of saccharide unit. Thus, $T_g$ of the sugar based surfactants are much higher than that of the other nonionic surfactants as mentioned above. That is, sugar-based surfactants possess a remarkable glass forming ability comparing to another type of surfactant. Because $T_g$ of anhydrous sugar-based surfactant existed almost above the freezing point of water, 0 °C, therefore, we expected that the behavior and ability of making glass state of the aqueous sugar-based surfactant solution can be readily observed without ice freezing if the cooling was conducted rapidly.

Recently, authors studied the vitrification or glassification of the aqueous solution of sugar-based surfactant, which must be associated with the specific function under freezing state (Ogawa, S. & Osanai, S. (2007). Ogawa, S.; Asakura, K. & Osanai, S. (2010).). In this chapter, we would like to elucidate some aspects of the aqueous solution of sugar-based surfactant under supercooling, where the simple primary phase transition such as gelation and crystallization is not a key topic, but the vitrification plays an important role. We would like to indicate the some characteristics of the sugar-based surfactant in an aqueous solution under low temperature. The basic behavior of these surfactants solution will be focused on the following items.

- The glass transition of the aqueous solution of sugar-based surfactant under low temperature with forming thermotropic and lyotropic liquid crystalline phases.
- Correlation between the glass transition and the protective effect against freezing.

The information obtained from this chapter would be valuable to the researchers who engage the low temperature technologies.

## 2. Glass transition behavior of octyl β-D-glucoside/water binary mixtures

Octyl β-D-glucoside (G8Glu: Scheme 1) is one of the representative sugar-based surfactants. Although there are many reports on phase behavior of C8Glu/water binary system (Boyd, B. J.; Drummond, C. J.; Krodkiewska, I. & Grieser, F. (2000). Nilsson, F.; Söderman, O. & Johansson, I. (1996). Häntzschel,D.; Schulte, J.; Enders, S. & Quitzsch, K. (1999). Dörfler, H.-D. & Göpfert, A. (1999). Bonicelli, M. G.; Ceccaroni, G. F. & La Mesa, C. (1998). Sakya, P., Seddon, J. M. & Templer, R. H. (1994). Loewenstein, A. & Igner, D. (1991). Kocherbitov, V.; Söderman O. & Wadsö, L. (2002).), no report was presented on its vitrification behavior under the low temperature. In this section, we introduce the glass transition behavior of octyl β-D-glucoside/water binary mixture within a wide concentration range under the conditions without ice formation (Ogawa, S.; Asakura, K. & Osanai, S. (2010).).

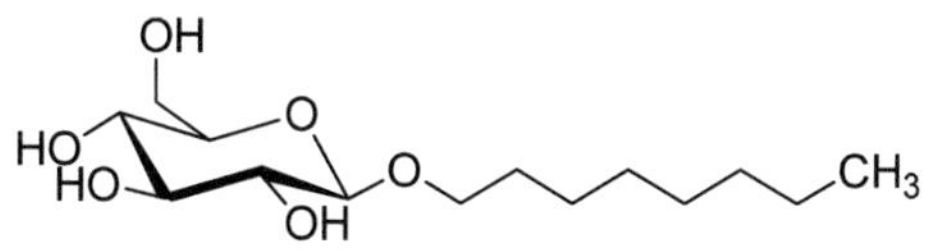


Scheme 1. Chemical structure of octyl β -D-glucoside (C8Glu).

C8Glu was synthesized as described in the literature, with a little modification (Bryan, M. C.; Plettenburg, O.; Sears, P.; Rabuka, D.; Wacowich-Sgarbi, S. & Wong, C.-H. (2002).).

### 2.1 Thermal behavior of C8Glu/water binary system

Fig. 1 shows a typical DSC chart which illustrates the glass transition behavior of G8Glu/water mixture. Each sample with various concentrations was homogenized by heating until 120 °C prior to the measurement. The sample was rapidly cooled to −120 °C at −10 °C/min and then heated at the rate of 10 °C/min. As Fig. 1 shows, when the concentration of C8Glu was greater than ca. 80 wt%, no ice was produced during cooling, and the glass transition was observed during the heating process. Occurrence of the glass transition was confirmed by the discontinuity of the heat capacity as indicated by solid line arrows in Fig. 1. In the concentration range from ca. 80 to 82 wt% for C8Glu, the ice was formed by the devitrification and thawed in the heating process (Fig. 1(a)). Devitrification was defined as the solidification phenomena after the temperature exceeded $T_g$ in the heating process.

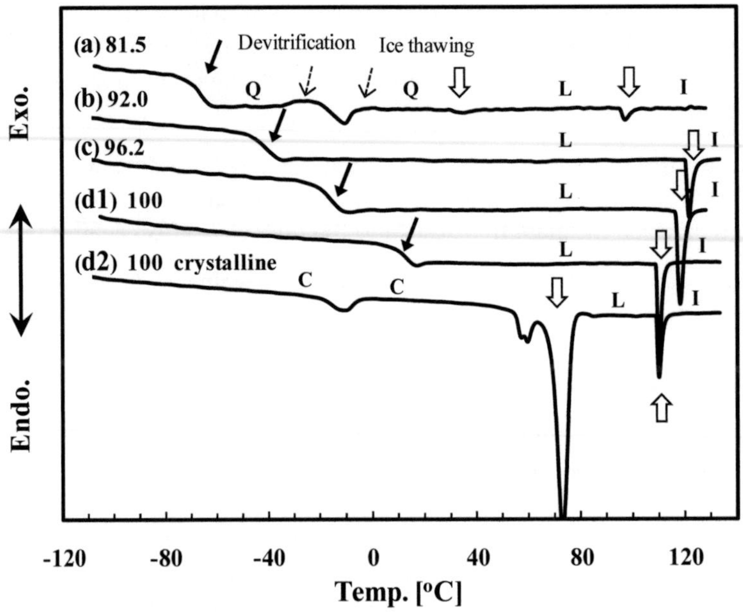


Fig. 1. Typical DSC thermograms of heating process at 10 °C/min of C8Glu/water mixture.

The concentration of the sample is expressed in wt %. White arrows indicate the phase transition between the liquid crystalline phases or from the liquid crystalline phases to an isotropic solution or melt. Solid line arrows indicate the glass transition, as mentioned above. I: isotropic solution, H: hexagonal phase, Q: cubic phase, L: lamellar phase, C: crystalline phase. *Apparatus*; DSC 60 (SHIMAZU Co. Ltd.) equipped with a cooling accessory was used throughout the measurement. *Sample preparation*; Samples of an aqueous solution were prepared from C8Glu and the prescribed amount of water. The sample was prepared as following two methods. *Method A*; A dilute aqueous C8Glu solution (ca. 35 wt%) in an aluminium pan was directly concentrated by drying over phosphorous pentoxide at an ambient temperature. *Method B*; A prescribed amount of water was absorbed under a humid atmosphere or added directly to C8Glu that was free from water. The anhydrous C8Glu was prepared by placing the sample on a hot stage at 125 °C for 50 min under a $N_2$ atmosphere to remove any water.

G8Glu/water binary system gave various kinds of liquid crystalline (LC) phase, such as hexagonal (H), cubic (Q), and lamella (L) phases and crystalline phase (C) according to its concentration and temperature. Fig. 1(b) indicates that L phase existed after the glass transition took place at around −40 °C during a heating process and it changed into isotropic liquid at 120 °C. In other words, the glass transition did not occur in crystalline and isotropic liquid phases but in a LC phase during the cooling process. A detailed comparison of (d1) with (d2) in Fig.1 clearly demonstrated that the phase transition from glass to lamella occurred in the LC phase, because the peak due to the transformation from crystalline to lamella LC at 70 °C was not observed in (d1) chart.

Observation by the polarizing optical microscopy (POM) gave a consistent result with these findings mentioned above. The sample was rapidly cooled to −100 °C at −10 °C/min and then heated at the rate of 10 °C/min. Fig. 2 shows the POM images of the C8Glu sample solution. Its conditions are shown in a legend of the figure. As can be seen from Fig. 2a and 2b, the L texture exhibited oily streaks at both temperatures above and below $T_g$. It indicated that L phase texture was maintained above or below $T_g$. It presented unambiguous evidence that this sample changed from liquid crystal to a glassy phase with holding its lamella structure, that is, the "glassy liquid crystals". The basic concept of the glassy liquid crystal was introduced in references. (Yoshioka, H.; Sorai, M. & Suga, H. (1983). Kocherbitov, V. & Söderman, O. (2004).

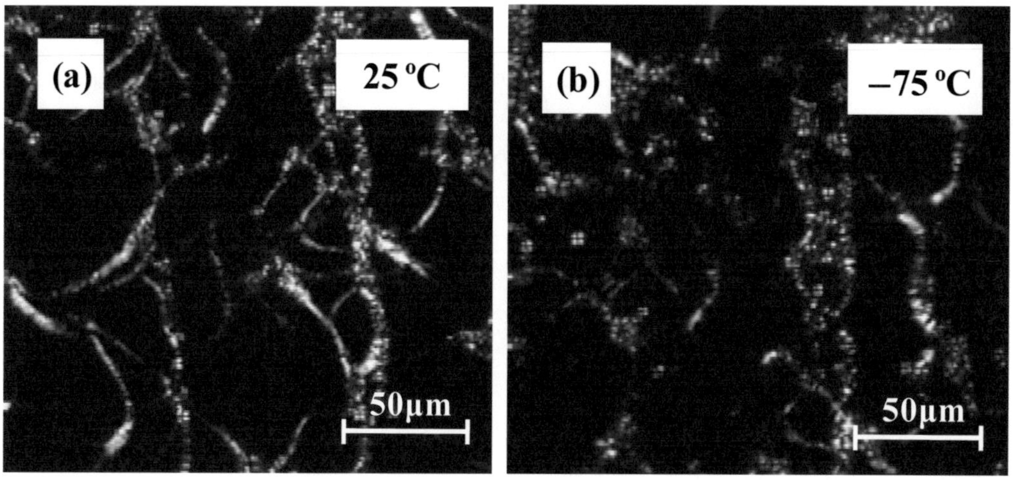


Fig. 2. POM photographs of 92.5 wt% of C8Glu/water mixture above and below $T_g$ of −47 °C.

*Apparatus*; Polarizing microscopy (BH-51, Olympus) equipped with a heating/cooling stage was used for observation. *Sample preparation*; Sample solutions were prepared in a similar manner to the DSC measurement. Sample was observed through a thin specimen sandwiched between a slide plate and cover-glass plate.

### 2.2 Concentration-temperature phase diagram with $T_g$ curve

Fig. 3 is a phase diagram of the C8Glu/water binary system at concentrations of more than 50 wt%. The diagram was constructed on the basis of experimental results obtained from DSC thermograms and POM photographs. Interested readers are able to refer detail methods for the determination of LC phases from numerous references as mentioned at the beginning of this section.

In this diagram, $T_g$ curve, the ice nucleation temperature curve (INC), and devitrification temperature curve (DC) are depicted. Although INC and DC curves are variable parameter according to the rate of nucleation, they are useful to understand the dynamic behavior of the system. As shown in Fig. 3, as the concentration of C8Glu increases, the lyotropic aggregates change from an isotropic solution (Micelle solution: M) to the liquid crystal phase, such as, H, Q, and L phases at 0 °C. When the concentration of C8Glu was lower than 80%, the INC was clearly recognized. It meant that the crystallization preferentially occurred below this temperature before the vitrification took place.

A lot of reference showed phase diagrams of C8Glu/water system that expressed the existence of crystal phase or gel phase in the concentrated region over 90 wt%. But in our study, the crystal phase did not appear in the same concentration at cooling rate −10 °C/min, between −120 °C and 120 °C. Instead of that, the glass transition was observed over ca. 80 wt% concentration. That was referred to $T_g$ curve in Fig. 3. The glass transition temperature, $T_g$, shifted to higher, as the concentration of C8Glu moved to higher. It means that C8Glu did not work as a plasticizer but as a curative agent in an aqueous solution.

Comparing the phase diagram above and below $T_g$ curve in Fig. 3, we are able to understand that the glassy phase was formed by cooling both of Q and L phases. It could therefore be presumed that, the formation of Q and L types of glassy LC phase occurred below $T_g$ curve. Even if temperature crossed the phase boundary between the Q and L phases, there was no discontinuity in $T_g$ line. It suggested that the difference among the liquid crystalline structures was not a decisive factor for the determination of $T_g$ in the aqueous solution.

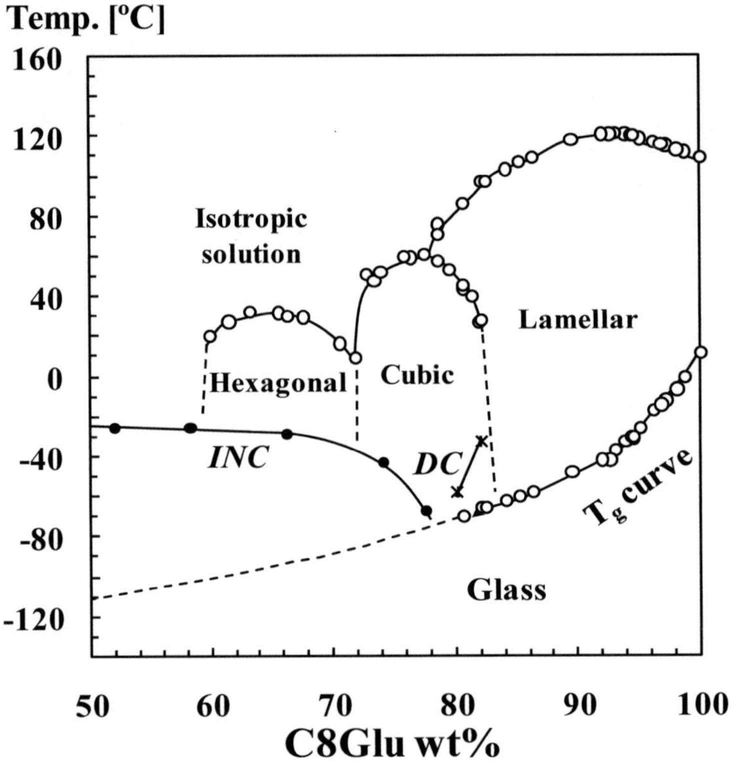


Fig. 3. Phase diagram of C8Glu/water mixture from 50 to 100 wt % C8Glu concentration including $T_g$ curve, ice nucleation curve (INC) and devitrification curve (DC).

The dotted lines are predicted one. The phase transition temperatures were determined by the intersection of the baseline and tangent to the end of the endothermic peak of the DSC chart on heating. $T_g$ was determined as the temperatures corresponding to half of the magnitude of the heat capacity change ($\Delta C_p$) at $T_g$ (Blond, G.; Simatos, D.; Catté, M.; Dussap,

C. G.; Gros, J. B. (1997).). The ice crystallization temperatures were defined as the average of five measurements.

### 2.3 Comparison of glass transition behavior with predicted curve

Couchman and Karasz presented a model that predicts $T_g$ of a mixture employing classical thermodynamics (Couchman, P. R. & Karasz, F. E. (1978). Couchman, P. R. (1978).). This model treated the glass transition as if it was equivalent to an Ehrenfest second order transition. The "original" Couchman-Karasz (C-K) formula is given below:

$$lnT_g = \frac{x_1 lnT_{g1} + kx_2 lnT_{g2}}{x_1 + kx_2} \tag{1}$$

where the subscripts 1 and 2 denote components 1 (C8Glu) and 2 (pure water), respectively. The symbols $x_1$ and $x_2$ represent the mole fractions of the corresponding C8Glu and pure water, respectively. $T_g$ is the glass transition temperature of the mixture under consideration; $k$ is a constant defined as $\Delta C_{p2}/\Delta C_{p1}$. $\Delta C_{p1}$ and $\Delta C_{p2}$ are the $\Delta C_p$ at $T_g$ of component 1, pure solute and component 2, pure water, respectively. Eq. 1 is often "modified" to the following general form:

$$T_g = \frac{x_1 T_{g1} + kx_2 T_{g2}}{x_1 + kx_2} \tag{2}$$

The suitability of these two equations ("original" and "modified" C-K equations) was discussed by comparison with the actual experimental measurements for $T_g$ of the G8Glu/water mixture.

Fig. 4 shows the $T_g$-prediction curves obtained from the "original" and "modified" C-K equations, using $\Delta C_{p1}$ = 142.2 J/mol K at $T_{g1}$ = 284.4 K (11.2 °C) for the amorphous C8Glu as obtained from our experimental results. At the same time, the experimental values were indicated in the corresponding figures. Values of $\Delta C_{p2}$ = 35.0 J/mol K and $T_{g2}$ = 135 K (−138.2 °C) for the pure water were taken from the literatures (Sugisaki, M.; Suga, H. & Seki, S. (1968). Rasmussen, D. H. & MacKenzie, A. P. (1971).). The experiment indicated that the analysis using the "original" (Eq. 1) gave a relatively good agreement with the experimental result over the entire concentration studied. By contrast, the predicted $T_g$ obtained from the "modified" (Eq. 2) (dotted line) was not in accord with the experimental finding. By considering these results, we can state the "original" C-K equation would give a much better prediction than the "modified" one for $T_g$ determination in mixtures. Couchman stated that if the assumption that $T_{g1}/T_{g2} \approx 1$ was applicable, the results obtained using the "modified" C-K would be valid (Couchman, P. R. & Karasz, F. E. (1978).). In our system, $T_{g1}$ of anhydrous C8Glu is 284.4 K, and giving a $T_{g1}/T_{g2}$ for C8Glu of 2.11. This value is far from the unity that is appropriate for the "modified" C-K equation. We guessed that the $T_g$ ratio between two components which form the mixture must be approximately one if we want to employ the 'modified' C-K equation to predict $T_g$.

$$\Delta C_p[J/gK] = \Delta F / (W_{sample} \times scan\ rate) \tag{3}$$

$$\Delta C_p[J/mol K] = \Delta C_p[J/gK](MW_1 \times x_1 + MW_2 \times x_2) \tag{4}$$

where $W_{sample}$ is the sample weight in a sealed pan and $\Delta F$ is the heat flow change (Fig. 4b). $MW_1$ and $MW_2$ are the molar weight of C8Glu [$MW_1$: 292.19], and pure water [$MW_2$: 18.02].

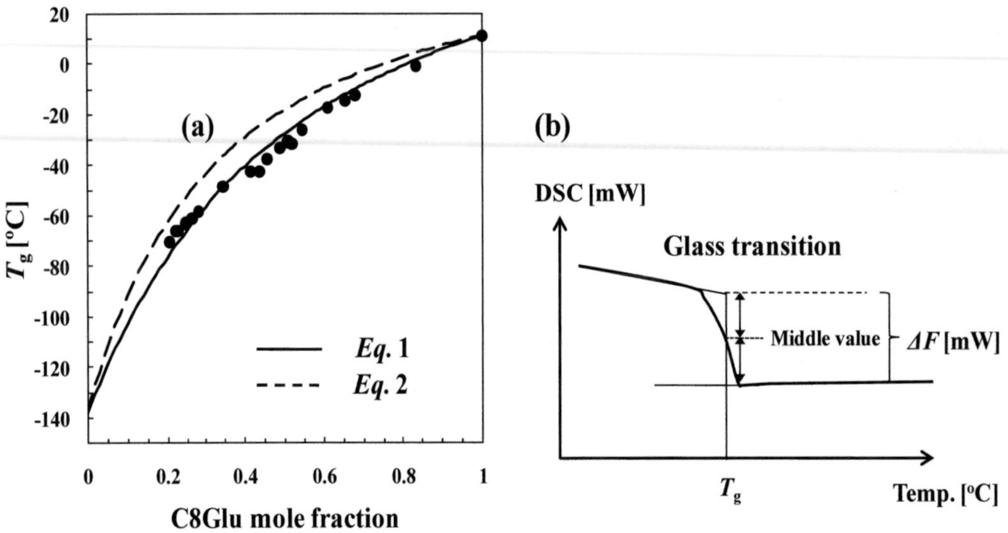


Fig. 4. (a) Comparison of experimental $T_g$ of C8Glu/water mixture obtained by two kinds of predicted curves by C-K equations and (b) estimation of $T_g$ and $\Delta F$.

Each plot shows the experimental data. $\Delta F$ was determined as sketched in Fig.4(b). Using this parameter, $\Delta C_p$ was obtained as follows.

### 2.4 Influence of concentration on $\Delta C_p$, $\Delta H$ and phase transition temperature $T_g$

Fig. 5 shows the $\Delta C_p$ curve as a function of the mole fraction of C8Glu. As mole fraction is decreasing from 1.0, there were two bending points at 0.65 and 0.40. The curve is divided into three regions (A, B, and C) according to their characteristics.

In the mixture of C8Glu/water, $\Delta C_p$ for pure water ($\Delta C_{p2}$) obtained by extrapolation of the plots in region C to zero C8Glu content, was 35.0 J/mol K. This value is compatible with the experimental value reported by Sugisaki et al. (Sugisaki, M.; Suga, H. & Seki, S. (1968).). On the other hand, $\Delta C_p$ for pure C8Glu ($\Delta C_{p1}$) obtained by extrapolation of the plots in region C to 1.0 C8Glu content, was about 175 J/mol K. There was an apparent difference between the extrapolated and actual experimental value, 142.2 J/mol K. By contrast, the extrapolation of plots in region A reached 0 J/mol K. These results showed that $\Delta C_p$ of the binary mixture was not predictable using a simple linear function composed of $\Delta C_{p1}$ and $\Delta C_{p2}$ .

In order to obtain information for clarification of this complex behavior of $\Delta C_p$, the transition between lamella and isotropic liquid phases were studied further in detail.

Fig. 6 indicates the relationships of the phase transition between lamella (L) and isotropic solution (I) with C8Glu mole fraction. Enthalpy ($\Delta H$) and temperature of the phase transition are depicted on the two vertical axes. Generally speaking, enthalpy of C8Glu solution decreased as the mole fraction of C8Glu reduced. A clear bending point was recognized at a particular concentration with mole fraction of 0.65, which was in fair agreement with that of the first bending point shown in Fig. 5. The enthalpy of the system reached 0 J/mol by extrapolation of plots in region A in analogy with the result shown in Fig. 5. These results meant that the amount of water in region A would have no influence not only on $\Delta C_p$ but also on ΔH.

Here, we adopted a concept of the "non-continuous water" to propose a hypothesis that interprets above behavior of $\Delta C_p$ in C8Glu/water mixture.

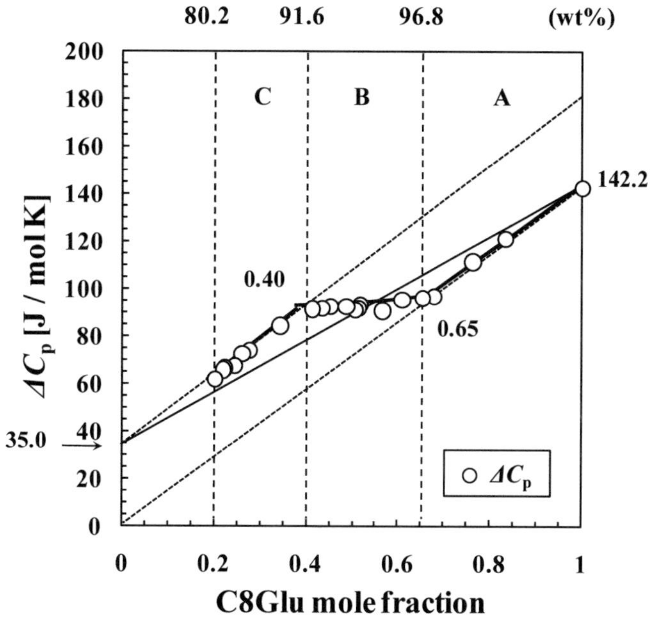


Fig. 5. $\Delta C_p$ behavior of C8Glu/water mixture with variation in concentrations.

The linear solid line connecting two $\Delta C_p$ values, 35.0 and 142.2 shows $\Delta C_p$ line when the mixing was carried out under holding ideal state.

We interpreted the behavior of C8Glu in different concentration of aqueous solution as follows: Fig. 7 shows the relationships among $\Delta C_p$ at $T_g$, $\Delta H$ of the phase transition and the phase transition temperature in C8Glu/water mixture systems on the basis of the experimental results. In region C, the molar ratio of water: C8Glu was 1.5 : 1 - 4 : 1, that is, molecule's number of water is larger than that of the C8Glu. The water in this region will constitute the aqueous phase keeping continuous state among a bimolecular membrane lamellar structure. It is a kind of bulk water. Reduction of water means simple decrease of the bulk water stated above. The fact that extrapolation of plots of $\Delta C_p$ in region C reached 35 J/mol K and coincided with that of pure water proved its validity.

On the other hand, in region A, the molar ratio of water : C8Glu is 1 : 2. In other words, the number of water molecule is less compared with that of C8Glu. The scarcity of water will be further signalized by consideration of their relative magnitude of the molecular bulkiness. This circumstance will not enable the water molecule to exist in continuous state. The water molecules in region A would be present in a non-continuous state with creating a new hydrogen bond among the glucoside molecules.

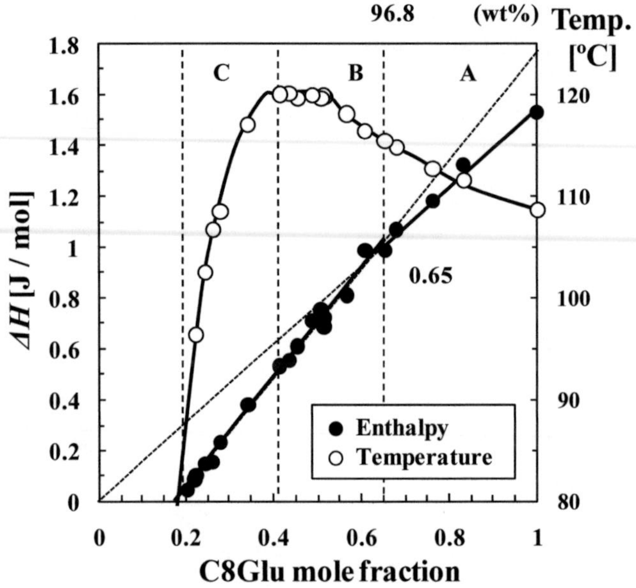


Fig. 6. Relationships of phase transition behavior between lamellar phase and isotropic solution with C8Glu fraction.

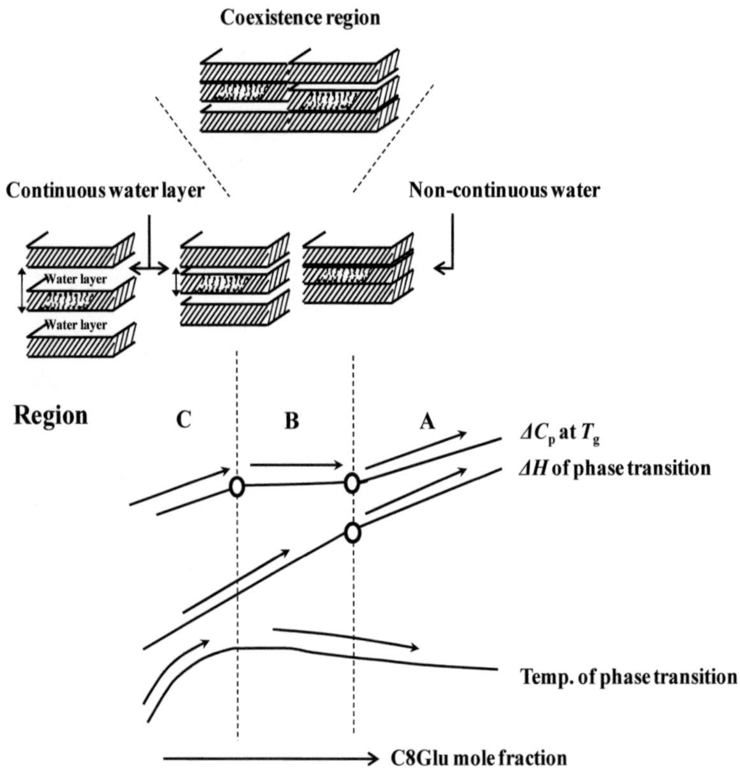


Fig. 7. Schematic figure of C8Glu/water mixture systems with variation of C8Glu concentration.

Region A: The water molecules would be present in a non-continuous state.

Region B: The water characteristic is determined by the mixed system composed of the continuous and non-continuous water existed in the region C and A.

Region C: The water behaves like bulk water and constitutes the aqueous phase keeping continuous state among a bimolecular membrane lamellar structure. Reduction of water means simple decrease of the bulk water.

In region B, $\Delta C_p$ kept in a constant state irrespective of its concentration of the system. This result demonstrated the additivity of two kinds of states at the corresponding concentration of the bending points in Fig. 5. It means the behavior of water in this region would be determined by the mixed system composed of the continuous and non-continuous water existed in the region C and A, respectively.

## 3. Glass transition behavior of octyl β-D-glucoside/NaCl/water ternary mixtures

The aqua-system of the life organism contains a various kinds of ions and exhibits complicated buffer actions to maintain its physiological functions in a normal state. As it is cooled down, eutectic phase composed of electrolyte and ice was generated in the concentrated unfrozen phase (Mullin, J. W. (2001).). Occurrence of the eutectic would be responsible of direct causes for damages against cells and enzymes and resulted unusual pH change would become a trigger for abnormal interactions (Heber, U.; Tyankova, L. & Santariu, K. A. (1971). Mollenhauer, A.; Schmitt, J. M.; Coughlan, S. & Heber U. (1983). Han, B. & Bischof, J. C. (2004). Wang, C.-L., Teo, K. Y. & Han. B. (2008). Goel, R.; Anderson, K.; Slaton, J.; Schmidlin, F.; Vercellotti, G.; Belcher, J. & Bischof, J. C. (2009).). In actual circumstances where the life organisms are treated under extremely and mildly cool atmosphere, various kinds of cryoprotectants and lyoprotectants such as salts, amino acids, carbohydrates, artificial and natural polymers are used to stabilize these bio-tissues from the cooling damages (Heber, U.; Tyankova, L. & Santariu, K. A. (1971). Tyankova, L. (1972). Izutsu, K.; Yoshioka, S. & Kojima, S. (1995). Koshimoto, C. & Mazur, P. (2002). Chen, N. J.; Morikawa, J. & Hashimoto, T. (2005). Chen, Y.-H. & Cui, Z. (2006). Kawai, K. & Suzuki, T. (2007). Izutsu, K.; Kadoya, S.; Yomota, C.; Kawanishi, T.; Yonemochi, E. & Terada, K. (2009).).

The purpose of this section is to clarify the inhibition effect of sugar-based amphiphiles on eutectic formation in the freeze-thawing process of aqueous NaCl solution (Ogawa, S. & Osanai, S. (2007).).

### 3.1 Thermal behavior of sugar-based amphiphiles/NaCl/water ternary system

Fig. 8 shows DSC charts of the ternary system consist of C8Glu/NaCl/water. A solution containing C8Glu at the same concentration (C8Glu to water = 1:9 [wt%]) was mixed with various concentration of NaCl as shown in Fig. 8. Each sample was cooled to −100 °C at −10 °C /min and then heated at the rate of 3 °C/min. The peak appeared at −21 °C referred to the fusion peak of eutectic of $NaCl \cdot 2H_2O$/ ice and another peak at about 0 °C was that of ice (Hvidt, A. & Borch, K. (1991).). These samples were classified into three groups according to the concentration of NaCl, Group I, II and III.

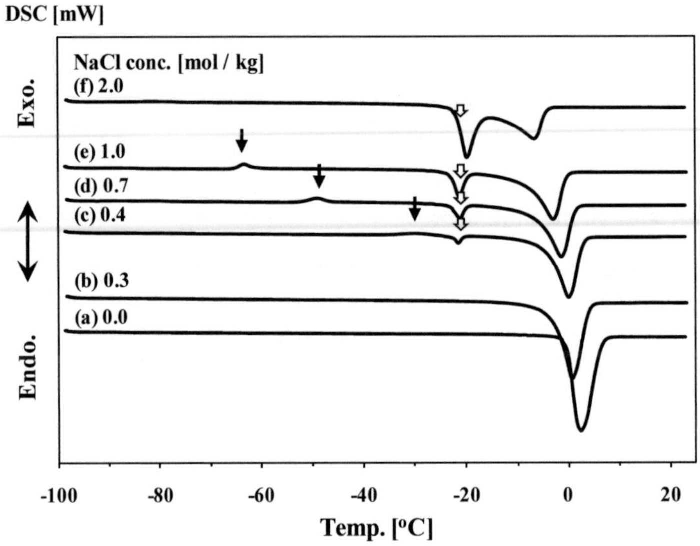


Fig. 8. DSC thermograms of C8Glu/NaCl/water systems in the thawing process at the heating rate of 3 °C/min.

The weight ratio between C8Glu and water was constant (C8Glu : water = 10 : 90 wt%). NaCl concentrations were shown in the figure. *Apparatus;* DSC 60 (SHIMAZU) was used throughout. *Sample preparation;* Each sample was prepared by dissolving NaCl and the C8Glu in a prescribed amount of water and leaving to stand for at least 2 h.

Group I: Chart (a) and (b) in Fig. 8. Their NaCl concentration was low. Only one peak due to fusion of ice was noticeable, that is, formation of the eutectic was completely restrained.

Group II: Chart (c), (d) and (e). Concentration of NaCl was moderate. The exothermic peak due to devitrification was also observed in addition to the peaks due to fusion of ice and eutectic were observed.

Group III: Chart (f). Concentration of NaCl was high. Only two peaks due to fusion of the eutectic and ice were observed and the devitrification was not recognized. The exothermic peak due to devitrification was also observed in addition to the peaks due to fusion of ice and eutectic.

### 3.2 Analysis of enthalpy for the fusion of eutectic and ice

Fig. 9a shows the relationship between the fusion enthalpy of the eutectic and the ice under the presence and absence of C8Glu. It was examined based on the each DSC chart in Fig. 8. In Fig. 9a, two dotted lines represent the corresponding results obtained under without C8Glu, that is, the result of NaCl solution. Quantitative analysis of the two peaks was conducted as shown in Fig. 9b.

As can be seen from Fig. 9a, it was confirmed that when the amphiphilic sugar derivative, C8Glu, was not present, the fusion enthalpy of ice decreased and that of the eutectic increased linearly with the concentration of NaCl. This result was interpreted that formation of the eutectic was regulated by NaCl concentration.

On the other hand, when C8Glu was present in the system at the concentration of 10 wt% of water mass, the fusion enthalpy of eutectic was zero in a region of Group (I), and that of the ice slowly decreased compared with that in other Groups. It meant that a part of water was retained as non-freezing water, which could not be attributed to the formation of ice even below −100 °C. In the Group (II) and (III) in Fig. 9a, dotted and solid two lines were depicted in parallel. It signified that formation of a definite amount of eutectic was depressed by C8Glu in the system regardless of NaCl concentration.

In this section, C8Glu clearly depicted the conception on the additive effect of amphiphilic sugar derivatives for eutectic formation. Some other sugar derivatives such as C12Raffinose, C12Sucrose, C12Maltose, C8Mannose, C8Gulose appeared in the following section also exhibited a similar behavior. From their nonspecific behavior, it was concluded that the characteristics that amphiphilic sugar derivatives possess the ability to depress the formation of eutectic was general one.

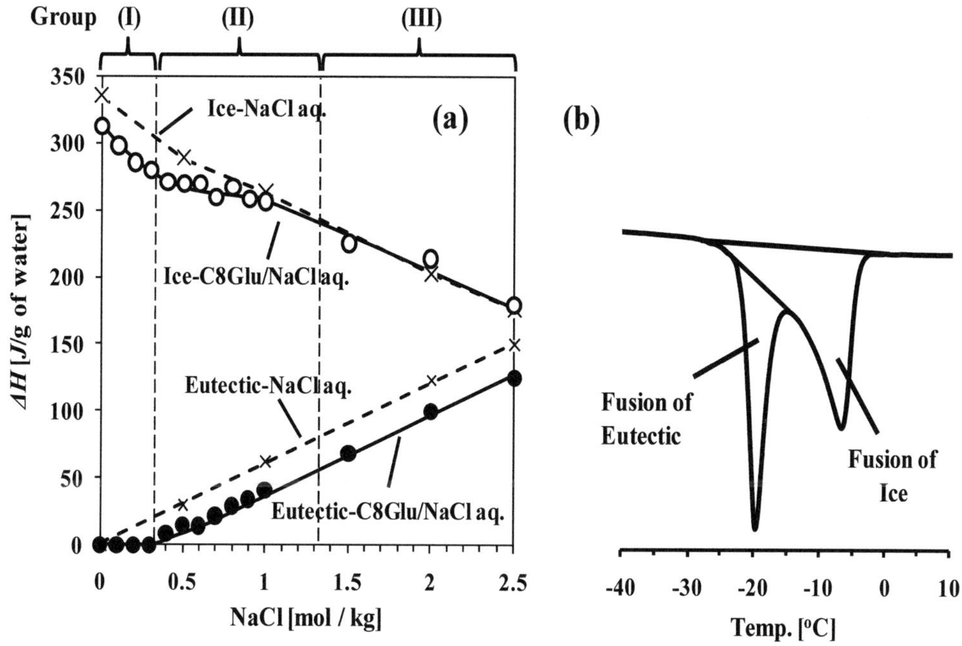


Fig. 9. Analysis of melting enthalpies in NaCl/water system with and without C8Glu.
(a) Fusion enthalpies of ice (above) and eutectic (below). (b) Calculated enthalpy areas of ice fusion and eutectic fusion.

### 3.3 Simultaneous XRD-DSC analysis

The depression effect of another amphiphilic sugar derivative for eutectic formation was studied to clarify its mechanism in detail. Here, C12Raf was used as a specimen instead of C8Glu. Scheme 2 shows its chemical structure and synthetic route.

Fig. 10 indicates DSC thermograms of C12Raf/NaCl/water ternary systems in the thawing process. The sample of C12Raf solution was prepared in a same concentration; C12Raf to water = 1:3 [weight ratio]. The molality of two NaCl solutions were 1.0 mol/kg and 2.5

mol/kg in Fig. 10(a) and (b). The appearance of the chart in Fig. 10(a) was similar to that of C8Glu system of Group (I) stated in Fig. 8 and Fig. 10(b) was to that of Group (II), respectively, although their sample situations were different in terms of their constituent and concentration.

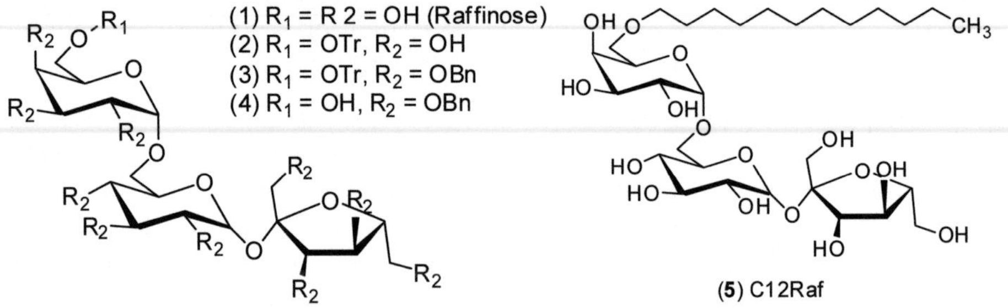


Scheme 2. Chemical structure and synthesis of 6″-*O*-dodecylraffinose (**5**: C12Raf).

C12Raf was synthesized from raffinose in four steps of tritylation, benzylation, detritylation, dodecylation and subsequent debenzylation, as shown in Scheme 2.

In two DSC charts in Fig. 10, an irregular deviation pointed out by an arrow was recognized on the base line. It appeared at −40 °C in (a) and −50 °C in (b). They were corresponding to a glass transition at this temperature, respectively. Fig. 10(a) suggested that the unfrozen phase was converted into the glass state after ice was built up during the cooling process. The exothermic peak appeared at around −40 °C during the heating process, in Fig. 10(b). It indicated that the devitrification conclusively occurred immediately after the glass transition. The unfrozen phase in a ternary sample became a glass state by freeze-condensation during a cooling process at −70 °C. Consequently, the formation of eutectic has been depressed under the kinetics.

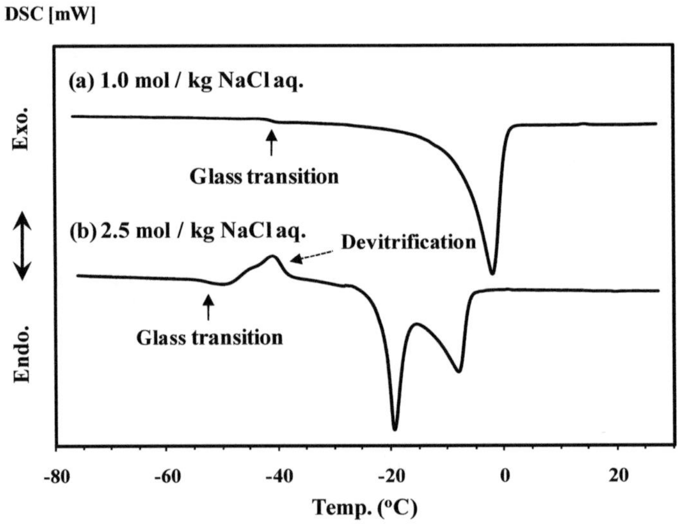


Fig. 10. DSC thermograms of C12Raf/NaCl/water systems in the thawing process.

The weight ratio between C12Raf and water was constant (C12Raf : water = 25 : 75 wt%). NaCl concentration was as follows; (a) 1.0 mol / kg of pure water and (b) 2.5 mol / kg of pure water.

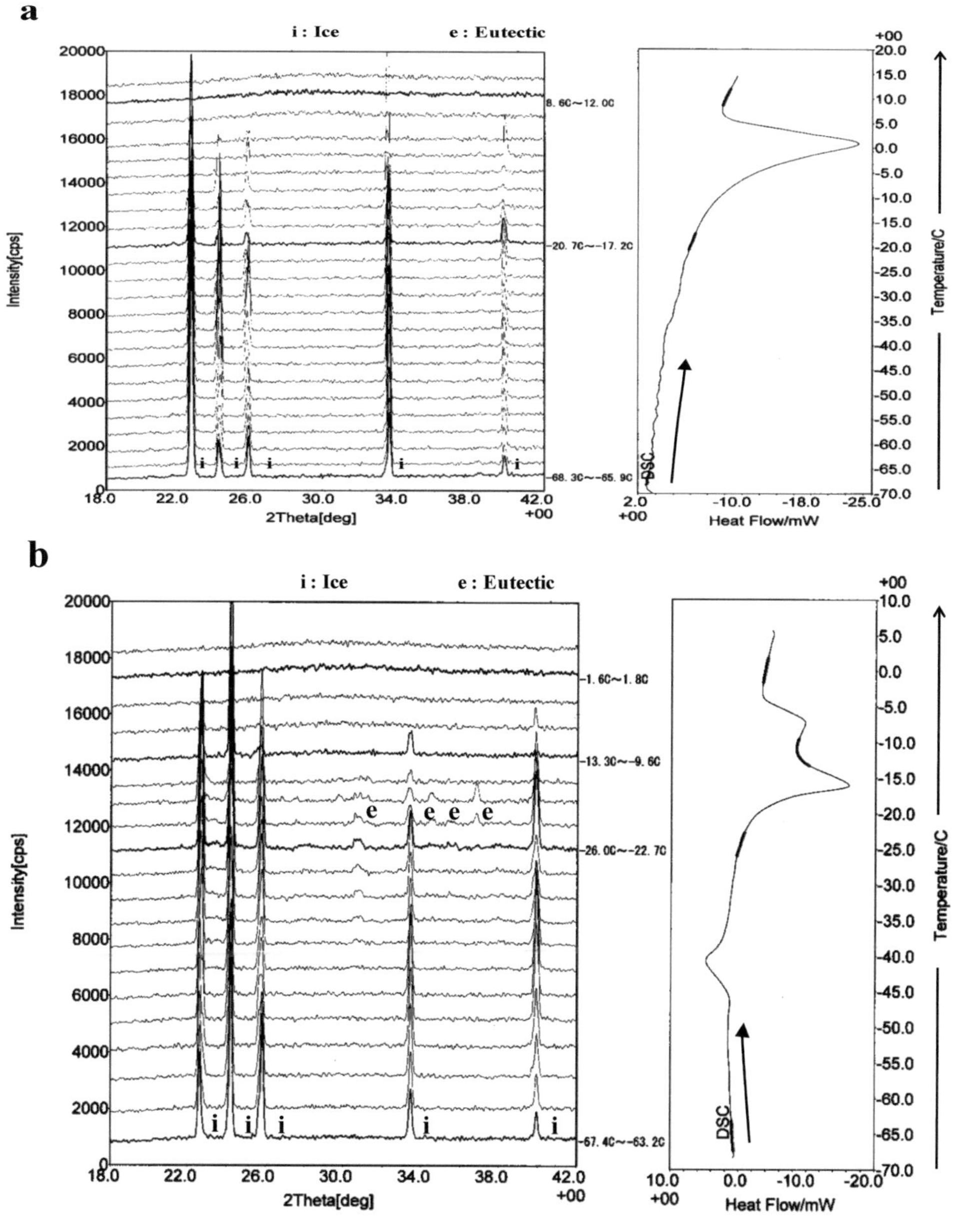


Fig. 11. Simultaneous XRD-DSC measurement of the thawing process in C12Raf/NaCl/water systems.

C12Raf and water was constant (C12Raf : water = 25 : 75 wt%). NaCl concentration was as follows; (a) 1.0 mol / kg of pure water and (b) 2.5 mol / kg of pure water. *Apparatus*; XRD-DSC II (RIGAKU) was used for measurement. Details of this apparatus are found elsewhere (Arii, T.; Kishi, A. & Kobayashi, Y. (1999). Kishi, A.; Otsuka, M. & Matsuda, Y. (2002).) *Measurement conditions* were as follows; *1*. Cooled to −70 °C at −6 °C/min. *2*. Heated to 15 °C at 2 °C/min. e, eutectic diffraction peak; i, ice diffraction peak.

The two samples in Fig. 10, (a) and (b) were examined by simultaneous XRD-DSC measurement. The results were summarized in Fig. 11. DSC chart of Fig. 11(a) showed only one peak due to the fusion of ice. The XRD-DSC chart demonstrated that when the system was cooled until −70 °C the ice was definitely formed. Five peaks at $2\theta$ = 22.5, 24.1, 25.8, 33.4, 39.8 [deg] were observed during the experiment. All diffraction peaks could be indexed to the standard hexagonal ice (Nishimoto, Y.; Kaneki, Y. & Kishi, A. (2004).). These peaks disappeared in the region above 0 °C. No peaks other than the ice were observed throughout the each and every temperature examined. It meant that formation of eutectic was completely depressed by C12Raf at this NaCl concentration.

Fig. 11(b) showed the XRD-DSC profiles for the sample prepared under a concentrated NaCl solution, its molality was 2.5 mol/kg. Highly meaningful results could be obtained by this method. In a cooler region of temperature between −67 °C ~ −30 °C, five peaks due to a hexagonal system of ice appeared at the same $2\theta$ angles as in Fig.11(a) in a similar manner. At higher temperature after an exothermic peak appeared at about −40.5 °C, four peaks newly emerged at $2\theta$ = 30.7, 34.5, 35.8, 36.8 [deg]. This peak pattern was in fair consistent with the authentic diffraction data of the eutectic, $NaCl \cdot 2H_2O$ / ice (Kajiwara, K.; Motegi, A. & Murase, N. (2001).). That is, it was found that the devitrification induced the formation of eutectic after the occurrence of the glass transition at −50 °C. These four peaks were extinguished accompanied by fusion of the eutectic above −21 °C. Further increment of temperature also resulted in a complete disappearance of the diffractive peaks of the ice.

These experiments were able to be summarized as follows; in a circumstance of dilute NaCl solution such as Group (I), the formation of eutectic was depressed by amphiphilic sugar derivatives such as C8Glu and C12Raf during both cooling and heating processes. On the other hand, in a medium concentrated NaCl solution designated Group (II), the formation of eutectic was restricted during the cooling process, but during the heating process, the devitrification induced the formation of eutectic after the occurrence of the glass transition. As could be seen from the Fig. 10, both in Group (I) and (II), the glass transition was confirmed during the heating process.

The glass formation plays a main role for this phenomenon, such as depression of eutectic formation. Non-amphiphilic free sugars and certain polymers have properties to change an aqueous solution into the glass state and inhibit the eutectic formation (Nicolajsen, H. & Hvidt, A. (1994). Izutsu, K.; Yoshioka, S. & Kojima, S. (1995). Kajiwara, K.; Motegi, A. & Murase, N. (2001).). The amphiphilic sugars would exhibit more effective capabilities except for depressing the formation of eutectic because of the versatile characteristics based on their interface active properties.

### 3.4 Effects of hydrophobic length and sugar structure on inhibition of eutectic formation

Two different kinds of sugars with a hydrophobic group or without it were examined to make clear the influence of the hydrophobic groups on the inhibition effect for eutectic formation. In other word, the effect of formation of aggregate of the specimen was examined. The results were summarized in Fig. 12. 6"-*O*-Dodecyraffinose (C12Raf) and 6'-*O*-dodecanoylsucrose (C12Suc) were used as specimens. The former linked the hydrophobic dodecyl group through ether linkage and the latter combined it through dodecanoyl ester linkage.

We separately confirmed their aggregation behavior using an automatic digital Kyowa Surface Tensiometer, CBVP-3 (Kyowa Kaimen Kagaku Ltd) by Wilhelmy-plate method. It was found that these amphiphilic sugar derivatives, C12Raf and C12Suc, had a critical micelle concentration (cmc) in pure water at 0.49 mM and 0.16 mM, respectively under room temperature. It meant these sugars formed the aggregate in the measuring conditions.

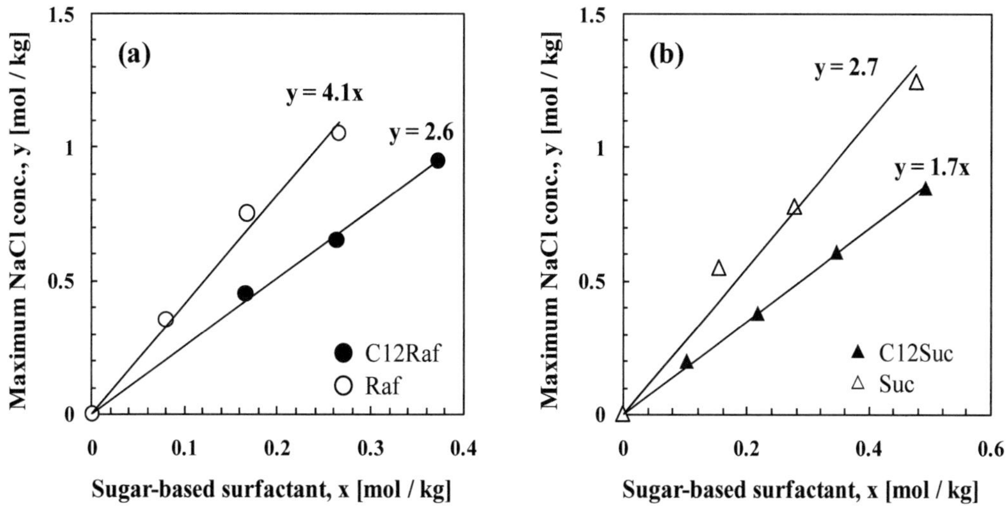


Fig. 12. NaCl concentration range of Group (I) with various concentrations of sugar-based amphiphiles. (a) C12Raf and Raf. (b) C12Suc and Suc.

Fig. 12 demonstrated relationships between the inhibited maximum NaCl concentration and the amphiphilic sugar concentration of the system in terms of two kinds of sugar. The ordinate suggested the maximum concentration of NaCl where the formation of eutectic was completely inhibited corresponding to the concentration of the sugar on the abscissa. This NaCl concentration refers to the boundary one between Group (I) and (II) shown in Fig. 9(a).

As can be seen from Fig. 12, depression ability for the formation of eutectic was clearly proportional to concentration of the sugar. The amphiphilic trisaccharide (C12Raf) and disaccharide (C12Suc) showed smaller depression ability than the corresponding non-amphiphilic free sugar. Its ratio was about 0.63 for all sugars examined.

The slope of the graph in Fig. 12 suggests the magnitude of the depression ability expressed in units per sugar molality. Fig. 13 showed the comparison of various kinds of sugars on the depression effect for eutectic formation. Amphiphilic glucose (C8Glu), mannose (C8Man) and gulose (C8Gul) are monosaccharide, sucrose (C12Suc) and maltose (C12Mal) are disaccharides, and raffinose (C12Raf) is trisaccharide. As can be seen from Fig. 13, the depression ability for the formation of eutectic of the sugar derivatives was proportional to the number of saccharide unit that constituted the hydrophilic part of the amphiphiles. The formation of eutectic made from about 0.8 ~ 0.9 molality of NaCl solution was inhibited by a unit molality of the sugar derivative per single unit of the saccharide in a proportional manner. In contrast to this, the epimeric isomerism and the structural isomerism between aldose and ketose gave little influence on the capability of inhibition of eutectic formation.

In Group (I) region, the depression effect for the eutectic formation resulted from the vitrification of an unfrozen aqueous phase during the cooling process. $T_g$ of anhydrous amphiphilic sugar derivatives of which the number of sugar unit are different were as follows: C8Glu =11.2 °C (Ogawa, S.; Asakura, K & Osanai, S. (2010).); C8Mal = 50.4 °C (Kocherbitov, V. & Söderman, O. (2004).; C12Maltotrioside = 100 °C (Ericsson, C. A.; Ericsson, L. C. & Ulvenlund, S. (2005).)). As can be seen from this, the $T_g$ of the sugar derivatives increased as the number of sugar unit increased. It was confirmed that the facility making vitrification was closely associated with the number of the sugar per a unit volume of the system or density of it.

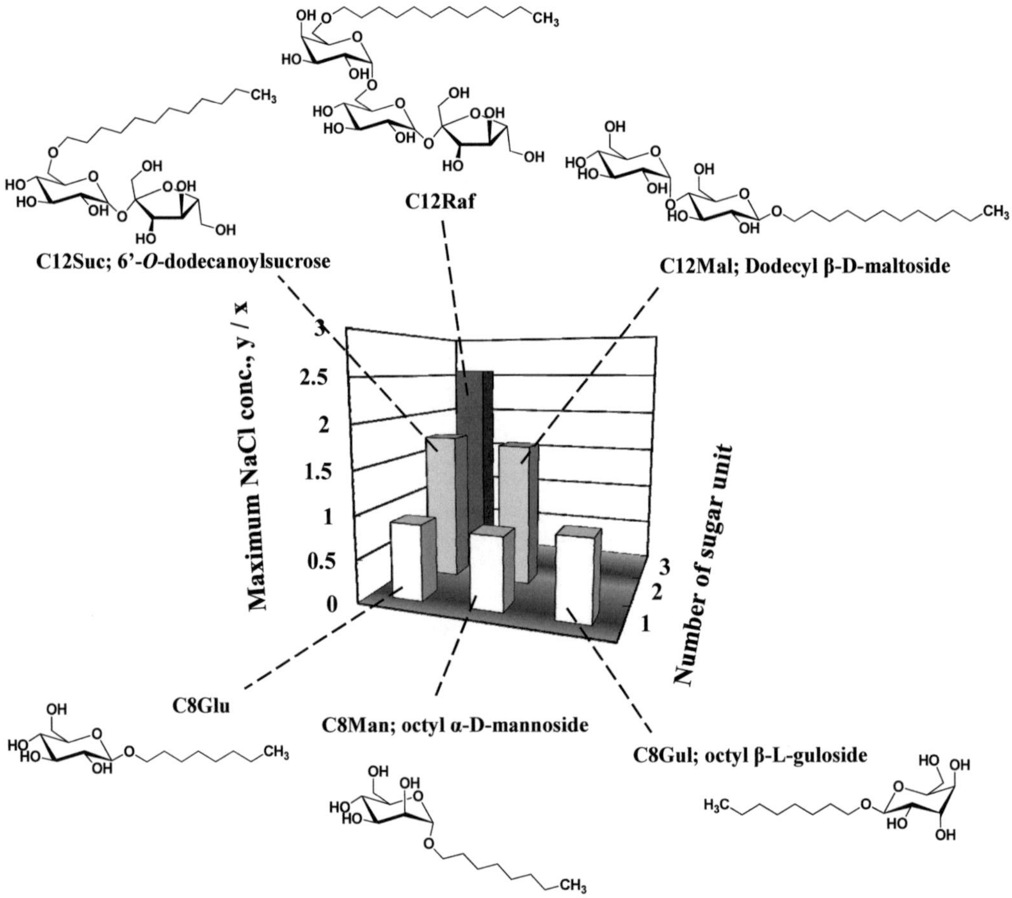


Fig. 13. Comparison of inhibition effect on eutectic formation with sugar structure.

*Material;* C12Suc, C8Man, C8Gul, and C12Mal were prepared according to published procedures (Ferrer, M.; Cruces, M. A.; Bernabé, M., Ballesteros, A. & Plou, F. J. (1999). Bryan, M. C.; Plettenburg, O.; Sears, P.; Rabuka, D.; Wacowich-Sgarbi, S. & Wong, C.-H. (2002). ).

## 4. Conclusion

Although sugar-based surfactants possess extraordinarily high $T_g$ in an anhydrous state, little is known about the actual application for its excellent glass forming ability. So far as we, authors know, this is the first attempt to apply it in aqueous system and description on it.

In this chapter, we presented the fundamental behavior of glass formation of sugar-based surfactant/water binary system and the inhibition effects of the sugar-based amphiphiles on the formation of eutectic that caused a lot of damage to a variety of bio-organisms from cells to proteins.

In Section 2, the outline of the glass transition behavior of C8Glu, which is one of the representative sugar-based surfactant, and water mixture system was described and summarized. It was clarified the formation of "lyotropic liquid crystal glass" generated from the liquid crystal such as cubic (Q) and lamella (L) in this system. The experimental data for $T_g$ of the lyotropic liquid crystal glass were in fair agreement with the theoretical values proposed as "original" equation by Couchman-Karasz. The peculiar behavior of the system observed through the change of specific heat ($\Delta C_p$) during the glass transition and enthalpy ($\Delta H$) of the phase transition from lamella to isotropic solution or fused liquid was discussed from the standpoint of the permeability of water molecule in the bimolecular membrane structure.

In Section 3, we mentioned the key aspects of the relationships between the inhibiting effect of the sugar-based surfactants and the generating of eutectic in the system. It was also confirmed that increasing saccharide unit of sugar-based surfactant induced an excellent inhibiting effect to the formation of eutectic. Although we focused the increment of the inhibiting power for the formation of eutectic on the introduction of hydrophobic group into the free sugar, the resulted sugar-based surfactant showed only 0.63 times ability for it comparing with the original free sugar.

Because the sugar-based surfactants possess not only the glass forming ability but also the interface active property in the same time, we could expect the possibility that these surfactants show some useful characteristic which could not be obtained by the ordinary free sugars. For example, various kinds of surfactant exhibit abilities that they can depress the deactivation of the protein during the freezing and thawing (Chang, B. S.; Kendrick, B. S. & Carpenter, J. F. (1996). Hillgren, A.; Lindgren, J. & Aldén, M. (2002).). But some surface active agents do not always show their contribution to maintain activities of the water soluble proteins such as LDH (Lactate Dehydrogenase) and β-Galactosidase in the freeze-drying treatment. In contrast to this, when a little amount of a certain sugar derivative was added to a system, it exhibited excellent effects for appreciable retention of the protein activities not only during freeze-thawing but also during freeze-drying processes (Izutsu, K.; Yoshioka, S. & Terao, T. (1993, 1994). Izutsu, K.; Yoshioka, S. & Kojima, S. (1995).).

It has been well known that carbohydrates or sugars are materials that can easily form glass state. (Dave, H.; Gao, F.; Lee. J.-H.; Liberatore, M.; Ho, C.-C. & Co, C. C. (2007).). The sugar-based surfactants could be considered as excellent multiple function surfactants, because they have two representative properties of the glass forming and the interface activity. Their application in an aqua-system expands its availability in the fields of foods, medicine and functional materials. Although they have the potential to play an advisable role, their application in a multicomponent system remains underdevelopment state. Under the current situation, the sugar-based surfactant has been applied in the bio-science fields, such as a preservation agent of proteins by freeze-drying method, a solubilizing agent for the preparation of reconstituted protein etc.

We expect that the research mentioned here would be further studied and contribute to their practical application of the sugar-based surfactants including the analytical development on the physico-chemical properties.

## 5. Acknowledgment

All of this study was carried out at "OleoScience Laboratory" in Faculty of Science and Technology, Keio University, Yokohama, JAPAN.

## 6. References

Amim, J.; Kawano, Y. & Petri, D. F. S. (2009). Thin Films of Carbohydrate Based Surfactants and Carboxymethylcellulose Acetate Butyrate Mixtures: Morphology and Thermal Behavior, *Materials Science and Engineering C-Biomimetic and Supramolecular systems*, Vol.C29, No.2, (March 2009), pp. 420-425 ISSN 0928-4931

Andersson, B. & Olofsson, G. (1987). Differential Scanning Calorimetric Study of Phase-Changes in Poly(ethylene glycol) Dodecyl Ether-Water Systems, *Colloid & Polymer Science*, Vol.265, No.4, pp.318-328, ISSN 030-402X

Arii, T.; Kishi, A. & Kobayashi, Y. (1999). A New Simultaneous Apparatus for X-ray Diffractometry and Differential Scanning Calorimetry (XRD-DSC), *Thermochimica Acta*, Vol.325, No.2, (January 1999), pp. 151-156, ISSN 0040-6031

Auvray, X.; Petipas, C. & Anthore, R. (1995). X-ray Diffraction Study of the Ordered Lyotropic Phases Formed by Sugar-Based Surfactants, *Langmuir*, Vol.11, No.2, (February 1995), pp. 433-439, ISSN 0743-7463

Bonicelli, M. G.; Ceccaroni, G. F. & La Mesa, C. (1998). Lyotropic and Thermotropic Behavior of Alkylglucosides and Related Compounds, *Colloid and Polymer Science*, Vol.276, No.2, (Feburary 1998), pp. 109-116, ISSN 0303-402X

Boyd, B. J.; Drummond, C. J.; Krodkiewska, I. & Grieser, F. (2000). How Chain Length, Headgroup Polymerization, and Anomeric Configuration Govern the Thermotropic and Lyotropic Liquid Crystalline Phase Behavior and the Air-Water Interfacial Adsorption of Glucose-Based Surfactants, *Langmuir*, Vol.16, No.19, (September 2000), pp. 7359-7367, ISSN 0743-7463

Blond, G.; Simatos, D.; Catté, M.; Dussap, C. G.; Gros, J. B. (1997). Modeling of the Water-Sucrose State Diagram Below 0 degrees C, *Carbohydrate Research*, Vol.298, No.3, (March 1997), pp. 139-145, ISSN 0008-6215

Bryan, M. C.; Plettenburg, O.; Sears, P.; Rabuka, D.; Wacowich-Sgarbi, S. & Wong, C.-H. (2002). Saccharide Display on Microtiter Plates, *Chemistry & Biology*, Vol.9, No.6, (June 2002), pp. 713-720, ISSN 1074-5521

Chang, B. S.; Kendrick, B. S. & Carpenter, J. F. (1996). Surface-Induced Denaturation of Proteins during Freezing and its Inhibition by Surfactants, *Journal of Pharmaceutical Sciences*, Vol.85, No.12, (December 1996), pp. 1325-1330, ISSN 0022-3549

Chen, N. J.; Morikawa, J. & Hashimoto, T. (2005). Effect of Amino Acids on the Eutectic Behavior of NaCl Solutions Studies by DSC, *Cryobiology*, Vol.50, No.3, (June 2005), pp. 264-272, ISSN 0011-2240

Chen, Y.-H. & Cui, Z. (2006). Effect of Salts on the Freezing Denaturation of Lactate Dehydrogenase, *Food and Bioproducts Processing*, Vol.84, No.C1, (March 2006), pp. 44-50, ISSN 0960-3085

Couchman, P. R. (1978). Compositional Variation of Glass-Transition Temperatures. 2. Application of Thermodynamic Theory to Compatible Polymer Blends, *Macromolecules*, Vol.11, No.6, (November 1978), pp. 1156-1161, ISSN 0024-9297

Couchman, P. R. & Karasz, F. E. (1978). Classical Thermodynamic Discussion of Effect of Composition on Glass-Transition Temperatures, *Macromolecules*, Vol.11, No.1, (January 1978), pp. 117-119, ISSN 0024-9297

Dave, H.; Gao, F.; Lee. J.-H.; Liberatore, M.; Ho, C.-C. & Co, C. C. (2007). Self-Assembly in Sugar-Oil Complex Glasses, Nature Materials, Vol.6, No.4, (April 2007), pp. 287-290, ISSN 1476-1122

Dörfler, H.-D. & Göpfert, A. (1999). Lyotropic Liquid Crystals in Binary Systems n-Alkyl Glycosides/Water, Journal of Dispersion Science and Technology, Vol.29, No.1-2, pp. 35-58, ISSN 0193-2691

Drummond, C. J.; Fong, C.; Krodkiewska, I.; Boyd, B. J. & Baker, I. J. A. (2003). 3 Sugar Fatty Acid Esters, In: *Novel Surfactants/114*, Edited by Holmberg, K., pp. 95-128, CRC Press, ISBN 978-0-203-91173-0, Marcel Dekker, New York

Ericsson, C. A.; Ericsson, L. C.; Kocherbitov, V.; Söderman, O. & Ulvenlund, S. (2005). Thermotropic Phase Behaviour of Long-chain Alkylmaltosides, *Physical Chemistry Chemical Physics*, Vol.7, No.15, (June 2005), pp. 2970-2977, ISSN 1463-9076

Ericsson, C. A.; Ericsson, L. C. & Ulvenlund, S. (2005). Solid-state Behaviour of Dodecylglycosides, *Carbohydrate Research*, Vol.340, No.8, (June 2005), pp. 1529-1537, ISSN 0008-6215

Ferrer, M.; Cruces, M. A.; Bernabé, M., Ballesteros, A. & Plou, F. J. (1999). Lipase-Catalyzed Regioselctive Acylation of Sucrose in Two-Solvent Mixtures, *Biotechnology and Bioengineering*, Vol.65, No.1, (October 1999), pp. 10-16, ISSN 0006-3592

Fong, C.; Weerawardena, A.; Sagnella, S. M.; Mulet, X.; Krodkiewska, I.; Chong, J. & Drummond, C. J. (2011). Monodisperse Noninonic Isoprenoid-Type Hexahydrofarnesyl Ethylene Oxide Surfactants: High Throughput Lyotropic Liquid Crystalline Phase Determination, Langmuir, Vol.27, No.6, (March 2011), pp. 2317-2326, ISSN 0743-7463

Fukada, K.; Matsuzaka, Y.; Fujii, M.; Kato, T. & Seimiya, T. (1998). Phase Behavior and Lyotropic-Liquid Crystal Structure of Alkyltrimethylammonium bromide-Water Mixtures around Freezing Temperature of Water, *Thermochimica Acta*, Vol.308, No.1-2, pp.159-164, ISSN 0040-6031

Gill, I. & Valivety, R. (2000a). Monosaccharide-Alkyl Glycoside Glass Phases: Plasticization with Hydrophilic and Hydrophobic Molecules, *Angewandte Chemie International Edition*, Vol.39, No.21, (November 2000), pp.3801-3804, ISSN 1433-7851

Gill, I. & Valivety, R. (2000b). Enzymatic Glycosylation in Plasticized Glass Phases: A Novel and Efficient Route to O-Glycosides, *Angewandte Chemie International Edition*, Vol.39, No.21, (November 2000), pp. 3804-3808., ISSN 1433-7851

Goel, R.; Anderson, K.; Slaton, J.; Schmidlin, F.; Vercellotti, G.; Belcher, J. & Bischof, J. C. (2009). Adjuvant Approaches to Enhance Cryosurgery, *Journal of Biomechanical Engineering*, Vol.131, No.7, (July 2009), pp. 074003-1-11, ISSN 0148-0731

Han, B. & Bischof, J. C. (2004). Direct Cell Injury Associated with Eutectic Crystallization During Freezing, *Cryobiology*, Vol.48, No.1, (February 2004), pp. 8-21, ISSN 0011-2240

Häntzschel, D.; Schulte, J.; Enders, S. & Quitzsch, K. (1999). Thermotropic and Lyotropic Properties of n-Alkyl-beta-D-Glucopyranoside Surfactants, *Physical Chemistry Chemical Physics*, Vol.1, No.5, (March 1999), pp. 895-904, ISSN 1463-9076

Hato, M.; Minamikawa, H. & Kato T. (2007). Chapter 10 Sugar-Based Surfactants with Isoprenoid-Type Hydrophobic Chains: Physicochemical and Biophysical Aspects, In: *Novel Surfactants/143*, Edited by Ruiz, C. C., pp. 361-411, CRC Press, ISBN 978-1-4200-5166-7, Taylor & Francis Group, Baco Raton

Heber, U. Tyankova, L. & Santariu, K. A. (1971). Stabilization and Inactivation of Biological Membranes during Freezing in the Presence of Amino Acids, *Biochmica et Biophysica Acta*, Vol.241, No.2, (August 1971), pp. 578-592, ISSN 0006-3002

Hill, K. LeHen-Ferrenbach, C. (2007). 1 Sugar-Based Surfactants for Consumer Products and Technical Applications, In: *Sugar-Based Surfactants/143*, Edited by Ruiz, C. C., pp. 1-20, CRC Press, ISBN 978-1-4200-5166-7, Taylor & Francis Group, Baco Raton

Hill, K. & Rhode, O. (1999). Sugar-Based Surfactants for Consumer Products and Technical Applications, *Fett/Lipid*, Vol.101, No.1, (January 1999), pp. 25-33, ISSN 0931-5985

Hillgren, A.; Lindgren, J. & Aldén, M. (2002). Protection Mechanism of Tween 80 During Freeze-Thawing of a Model Protein, LDH, *International Journal of Pharmaceutics*, Vol.237, No.1-2, (April 2002), pp.57-69, ISSN 0378-5173

Hoffmann, B.; Milius, W.; Voss, G.; Wunschel, M.; van Smaalen, S.; Diele, S. & Platz G. (2000). Crystal Structures and Thermotropic Properties of Alkyl α-D-glucopyranosides and Their Hydrates, Carbohydrate Research, Vol.323, No.1-4, (January 2000), pp. 192-201, ISSN 0008-6215

Hoffmann, B. & Platz, G. (2001). Phase and Aggregation Behaviour of Alkylglycosides, *Current Opinion in Colloid & Interface Science*, Vol.6, No.2, (April 2001), pp.171-177, ISSN 1359-0294

Hvidt, A. & Borch, K. (1991). NaCl-$H_2O$ Systems at Temperatures Below 273 K, Studied by Differential Scanning Calorimetry, *Thermochimica Acta*, Vol.175, No.1, (February 1991), pp. 53-58 ISSN 0040-6031

Imura T.; Hikosaka, Y.; Worakitkanchanakul, W.; Sakai, H.; Abe, M.; Konishi, M.; Minamikawa, H. & Kitamoto, D. (2007). Aqueous-Phase Behavior of Natural Glycolipid Biosurfactant Mannosylerythritol Lipid A: Sponge, Cubic, and Lamellar Phases, *Langmuir*, Vol.23, No4, (February) pp.1659-1663, ISSN 0743-7463

Izutsu, K.; Kadoya, S.; Yomota, C.; Kawanishi, T.; Yonemochi, E. & Terada, K. (2009). Freeze-Drying of Proteins in Glass Solids Formed by Basic Amino Acids and Dicarboxylic Acids, *Chemical & Pharmaceutical Bulletin*, Vol.57, No.1, (January 2009), pp.43-48, ISSN 0009-2363

Izutsu, K.; Yoshioka, S. & Terao, T. (1994). Stabilizing Effect of Amphiphilic Excipients on the Freeze-Thawing and Freeze-Drying of Lactate Dehydrogenase, *Biotechnology and Bioengineering*, Vol.43, No.11, (May 1994), pp. 1102-1107, ISSN 0006-3592

Izutsu, K.; Yoshioka, S. & Terao T. (1993). Stabilization of β-galactosidase by Amphiphilic Additives During Freeze-Drying, *International Journal of Pharmaceutics*, Vol.90, No.3, (March 1993), pp. 187-194, ISSN 0378-5173

Izutsu, K.; Yoshioka, S. & Kojima, S. (1995). Effect of Cryoprotectants on the Eutectic Crystallization of NaCl in Frozen Solutions Studied by Differential Scanning Calorimetry (DSC) and Broad-Line Pulsed NMR, *Chemical Pharmaceutical Bulletin*, Vol.43, No.10, (October 1995), pp. 1804-1806, ISSN 0009-2363

Izutsu, K.; Yoshioka, S. & Kojima, S. (1995). Increased Stabilizing Effects of Amphiphilic Excipients on Freeze-Drying of Lactate Dehydrogenase (LDH) by Dispersion into Sugar Matrices, *Pharmaceutical Research*, Vol.12, No.6, (June 1995), pp.838-843, ISSN 0724-8741

Jensen, R. E.; O'Brien, E.; Wang, J.; Bryant, J.; Ward, T. C.; James, L. T. & Lewis, D. A. (1998). Characterization of Epoxy-Surfactant Interactions, *Journal of Polymer Science Part B: Polymer Physics*, Vol.36, No.15, (November 1998), pp. 2781-2792, ISSN 0887-6266

Kajiwara, K.; Motegi, A. & Murase, N. (2001). Freeze-Thawing Behaviour of Highly Concentrated Aqueous Alkali Chloride-Glucose Systems, *CryoLetters*, Vol.22, No.5, (September-October 2001), pp. 311-320, ISSN 0143-2044

Kawai, K. & Suzuki, T. (2007). Stabilizing Effect of Four Types of Disaccharide on the Enzymatic Activity of Freeze-Dried Lactate Dehydrogenase: Step by Step Evaluation from Freezing to Storage, *Pharmaceutical Research*, Vol.24, No.10, (October 2007), pp. 1883-1890, ISSN 0724-8741

Kishi, A.; Otsuka, M. & Matsuda, Y. (2002). The Effect of Humidity on Dehydration Behavior of Nitrofurantoin Monohydrate Studied by Humidity Controlled Simultaneous Instrument for X-ray Diffractometry and Differential Scanning Calorimetry (XRD-DSC), *Colloids and Surfaces B: Biointerfaces*, Vol.25, No.4, (August 2002), pp. 281-291, ISSN 0927-7765

Kocherbitov, V. & Söderman, O. (2004). Glassy Crystalline State and Water Sorption of Alkyl Maltosides, *Langmuir*, Vol.20, No.8, (April 2004), pp. 3056-3061, ISSN 0743-7463

Kocherbitov, V. & Söderman, O. (2003). Phase Diagram and Physicochemical Properties of the n-Octyl α-D-glucoside/Water System, *Physical Chemical Chemical Physics*, Vol.5, No.23, (December 2003), pp.5262-5270, ISSN 1463-9076

Kocherbitov, V.; Söderman O. & Wadsö, L. (2002). Phase Diagram and Thermodynamics of the n-Octyl beta-D-Glucoside/Water System, *Journal of Physical Chemistry B*, Vol.106, No.11, (March 2002) pp. 2910-2917, ISSN 1520-6106

Koshimoto, C. & Mazur, P. (2002). The Effect of the Osmolality of Sugar-Containing Media, the Type of Sugar, and the Mass and Molar Concentration of Sugar on the Survival of Frozen-Thawed Mouse Sperm, *Cryobiology*, Vol.45, No.1, (August), pp. 80-90, ISSN 0011-2240

Loewenstein, A. & Igner, D. (1991). Deuterium NMR-Studies of n-Octyl alpha and beta-D-Glucopyranoside Liquid Crystalline Systems, *Liquid Crystals*, Vol.10, No.4, (October 1991), pp. 457-466, ISSN 0267-8292

Mollenhauer, A.; Schmitt, J. M.; Coughlan, S. & Heber U. (1983). Loss of Membrane-Proteins from Thylakoids during Freezing, *Bichimica et Biophysica Acta*, Vol.728, No.3, (March 1983), pp. 331-338, ISSN 0006-3002

Mullin, J. W. (2001). 4, Phase Equilibria, in *Crystallization Fourth Edition*, ISBN 0-7506-4833-3, Butterworth-Heinemann, Ocford.

Nibu, Y. & Inoue, T. (1998a). Solid-Liquid Phase Behavior of Binary Mixture of Tetraethylene Glycol Decyl Ether and Water, *Journal of Colloid and Interface Science*, Vol.205, No.2, (September 1998a), pp.231-240, 0021-9797

Nibu, Y. & Inoue, T. (1998b). Phase Behavior of Aqueous Mixtures of Some Polyethylene Glycol Decyl Ethers Revealed by DSC and FT-IR Measurements, *Journal of Colloid and Interface Science*, Vol.205, No.2, (September 1995), pp.305-315, ISSN 0021-9797

Nibu, Y.; Suemori, T. & Inoue T. (1997). Phase Behavior of Binary Mixture of Heptaethylene Glycol Decyl Ether and Water: Formation of Phase Compound in Solid Phase, *Journal of Colloid and Interface Science*, Vol. 191, No.1, (July 1997), pp. 256-263, ISSN 0021-9797

Nicolajsen, H. & Hvidt, A. (1994). Phase Behavior of the System Trehalose-NaCl-Water, *Cryobiology*, Vol.31, No.2, (April 1994), pp. 199-205, ISSN 0011-2240

Nilsson, F.; Söderman, O. & Johansson, I. (1996). Physical-Chemical Properties of the n-Octyl beta-D-Glucoside/Water System. A Phase Diagram, Self-Diffusion NMR, and SAXS Study, Langmuir, Vol.12, No.4, (February 1996), pp. 902-908, ISSN 0743-7463

Nishimoto, Y.; Kaneki, Y. & Kishi, A. (2004). Simultaneous XRD-DSC Measurements of Water-2-Propanol at Sub-Zero Temperatures, *Analytical Sciences*, Vol.20, No.7, (July 2004), pp. 1079-1082, ISSN 0910-6340

Ogawa, S.; Asakura, K. & Osanai, S. (2010). Glass Transition Behavior of Octyl beta-D-Glucoside and Octyl beta-D-Thioglucoside/Water Binary Mixtures, *Carbohydrate Research*, Vo.345, No.17, (November 2010), pp. 2534-2541, ISSN 0008-6215

Ogawa, S. & Osanai, S. (2007). Inhibition Effect of Sugar-based Amphiphiles on Eutectic Formation in the Freezing-Thawing Process of Aqueous NaCl Solution, *Cryobiology*, Vol.54, No.2, (April 2007), pp. 173-180, ISSN 0011-2240

Rasmussen, D. H. & MacKenzie, A. P. (1971). Glass Transition in Amorphous Water-Application of Measurements to Problems Arising in Cryobiology, *Journal of Physical Chemistry*, Vol.75, No.7, (April 1971), pp. 967-973, ISSN 0022-3654

Rybinski, von W. & Hill, K., Alkyl Polyglycosides-Properties and Applications of a new Class of Surfactants, *Angewandte Chemie International Edition*, Vol.37, No.10, (June 1998) pp.1328-1345, ISSN 1433-7851

Sakya, P., Seddon, J. M. & Templer, R. H. (1994). Lyotropic Phase Behavior of n-Octyl-1-O-beta-D-Glucopyranoside and its Thio Derivative n-Octyl-1-S-beta-D-Glucopyranoside, *Journal De Physiqu II*, Vol.4, No.8, (August 1994), pp. 1311-1331, ISSN 1155-4312

Söderberg, I.; Drummond, C. J.; Furlong, D. N.; Godkin, S. & Matthews, B. (1995). Non-ionic Sugar-Based Surfactants: Self Assembly and Air/Water Interfacial Activity, *Colloids and Surfaces A: Physicochemical and Engineering Aspects*, Vol. 102, (September 1995), pp.91-97, ISSN 0927-7757

Sugisaki, M.; Suga, H. & Seki, S. (1968). Calorimetric Study of the Glassy State. IV. Heat Capacities of Glassy Water and Cubic Ice, *Bulletin of the Chemical Society of Japan*, Vol.41, No.11, (November 1968), pp. 2591-2599, ISSN 0009-2673

Szűts, A.; Pallagi, E.; Regdon, G. Jr; Aigner, Z.; Szabó-Révész, P. (2007). Study of Thermal Behaviour of Sugar Esters, *International Journal of Pharmaceutics*, Vo. 336, No.2, (May 2007), pp. 199-27, ISSN 0378-5173

Tyankova, L. (1972). Effect of Amino-Acids on Thylakoid Membranes During Freezing as Influenced by Side Chain and Position on the Amino Group, *Biochimica et Biophysica Acta*, Vo.274, No.1, (July 1972), pp. 75-82, ISSN 0006-3002

Wang, C.-L., Teo, K. Y. & Han. B. (2008). An Amino Acidic Adjuvant to Augment Cryoinjury of MCF-7 Breast Cancer Cells, *Cryobiology*, Vol.57, No.1, (August 2008), pp. 52-59, ISSN 0011-2240

Warr, G. G.; Drummond, C. J.; Grieser, F.; Ninham, B. W. & Evans, D. F. (1986). Aqueous-Solution Properties of Nonionic n-Dodecyl Beta-D-Maltoside Micelles, *Journal of Physical Chemistry*, Vol.90, No.19, (September 1986), pp. 4581-4586, ISSN 0022-3654

Yoshioka, H.; Sorai, M. & Suga, H. (1983). Heat Capacity of N-p-n-Hexyloxybenzylidene-p'-n-butylaniline between 11 and 393 K: Unusual Glassy Smectic Liquid Crystal, *Molecular Crystal & Liquid Crystal*, Vol.95, No.1-2, (November 1982), pp. 11-30, ISSN 0140-6566

Zheng, L. Q.; Suzuki, M. & Inoue, T. (2002). Phase Behavior of an Aqueous Mixture of Octaethylene Glycol Dodecyl Ether Revealed by DSC, FT-IR, and 13C NMR Measurements, Langmuir, Vol.18, No.6, (March 2002), pp.1991-1998, ISSN 0743-7463

Zheng, L. Q.; Suzuki, M.; Inoue, T. & Lindman, B. (2002). Aqueous Phase Behavior of Hexaethylene Glycol Dodecyl Ether Studied by Differential Scanning Calorimetry, Fourier Transform Infrared Spectroscopy, and 13C NMR Spectroscopy, *Langmuir*, Vol.18, No.24, (November 2002), pp.9204-9210, ISSN 0743-7463

# 4

# Suppressing Method of Supercooling State in Cool Box Using Membrane

Seiji Okawa
*Tokyo Institute of Technology*
*Japan*

## 1. Introduction

### 1.1 Usage of cool box

In the area of food transportation, demand for cool boxes is increasing. There are many types of cool box which differ mainly by changing the melting temperature of refrigerant. Fixing of the melting temperature can be designed by arranging the composition of water and refrigerant and its concentration. In general, inorganic salt solutions such as sodium chloride, ammonium chloride or magnesium chloride are selected as the base and alcoholic solution such as methanol or ethanol and gelatinizing agent are added. To simplify the phenomena, sodium chloride solution is used in this research. Refrigerant in the cool box is necessary to be frozen before the use in order to store the latent heat. The stored energy is much bigger than the sensible heat. Growing demand for low freezing temperature use is rising. However, it is difficult to meet the need because of the existence of supercooling phenomenon.

### 1.2 Natural freezing with supercooling

When the solution is cooled down below the freezing temperature, liquid state remains which is called the supercooling state. The difference between the temperature when freezing occurred and the melting temperature is called the degree of supercooling at freezing. It is a statistical phenomenon and so it varies even the conditions are identical. It also depends on the volume and the cooling rate.

Fig. 1 shows one of the examples of natural freezing using sodium chloride solutions. Volume of solution is 0.1 ml and the cooling rate is 0.25 K/min. 20 experiments were carried out for each concentration and the average values were plotted.

Table 1 shows the melting temperature for each concentration of the solution (JSME, 1983). For example, for 10 wt% solution, the solution needs to be cooled down to -23.3 °C to obtain solidification. In a case of the cool box, since the volume and the cooling rate are different from the experimental conditions above, it needs to be installed in refrigerator having around 10 K below its melting temperature. The majority of domestic refrigerators only cool down to -18 °C. So consequently, one having melting temperature of around -10 °C has been a majority among all. Moreover, freezing of supercooled liquid is a statistical phenomenon, and so it needs around two days to assure the solidification. On the other hand, the demand for cool box with lower melting temperature is increasing and it leads to the necessity of installing another refrigerator with higher capacity.

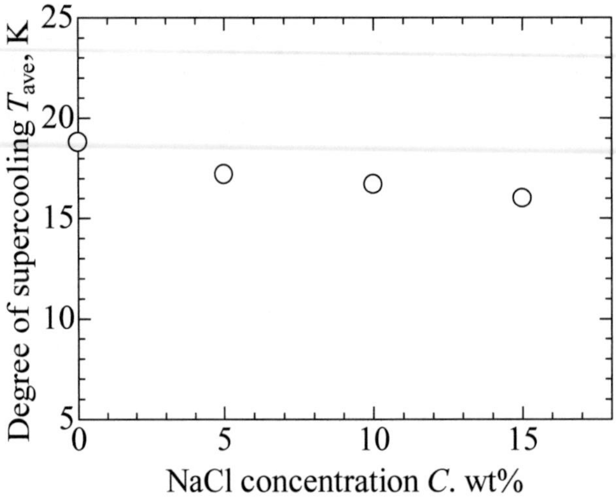


Fig. 1. Average degree of supercooling at freezing versus concentration of NaCl solution

| Concentration of NaCl | melting temperature | average degree of supercooling at freezing | average temperature for freezing |
|---|---|---|---|
| 0 wt% | 0.0 °C | 18.8 K | -18.8 °C |
| 5 wt% | -3.046 °C | 17.2 K | -20.3 °C |
| 10 wt% | -6.564 °C | 16.7 K | -23.3 °C |
| 15 wt% | -10.888 °C | 16.0 K | -26.9 °C |

Table 1. Melting temperature of NaCl concentrations and its average temperature for freezing (JSME, 1983)

In usual thermal cycle, such as a power plant, high efficiency can be obtained when there is a big temperature difference between the high temperature reservoir and the low temperature reservoir. So a lot of kinetic power can be obtained. In a case of refrigerant cycle, the situation is opposite. In order to obtain a big temperature difference, a high energy is required. Therefore, to obtain the low temperature, COP (coefficient of performance) becomes low. A lot of work needs to be done by a compressor to obtain the low temperature. It increases not only the cost of installation but also the electrical load consumption. Therefore, it is necessary to find a method to induce solidification of supercooled refrigerant.

### 1.3 Various methods to induce solidification of supercooled solutions

There are many researchers performed experiments to induce solidification. Collision or rubbing of solids or liquid in supercooled liquid induce solidification (Saito et al., 1992). Electrical charge on solidification of supercooled liquid is effective (Shichiri & Araki, 1986; Okawa et al, 1997, 1999; Hozumi et al, 2005). Applying ultrasonic wave to induce freezing of supercooled liquid is effective (Hozumi et al., 1999, 2002a, 2002b; Inada et al., 2001). A method to predict the degree of supercooling when Silver Iodide particle is used as nuclei was introduced (Okawa et al., 2001).

### 1.4 Purpose

A capsule having a wall made of ion exchange membrane and containing water inside was invented (Okawa et al., 2010a, 2010b), so only water can go through it. By installing the capsule in cool box, water in a capsule freeze first because of higher melting temperature and the membrane becomes a trigger for refrigerant to freeze with very low degree of supercooling. Refrigerant package as cool box consists of a thermal storage material which has low melting temperature. By installing similar liquid material which has higher melting temperature into the refrigerant package, the liquid material with higher melting temperature freezes first and it becomes a trigger for refrigerant to freeze. Liquid material with higher melting temperature in a capsule is isolated from refrigerant by using a membrane to separate with. Hence only water can go through between the liquid material and refrigerant. The purpose of this research is to clarify the ice propagation phenomena using 3 different kinds of membranes, experimentally.

## 2. Type of membrane

### 2.1 Ion exchange membrane

There are many types of membranes used for liquids. Microfiltration membrane is for eliminating microorganisms or particles having the size around 0.1 µm to 1 µm. Uultrafiltration membrane is for eliminating particles or polymers having the size around 2 nm to 0.1 µm. Nanofiltration membrane is for eliminating particles or polymers having the size smaller than 2 nm. Reverse osmosis membrane is for desalination of sea water and waste water treatment. Dialysis membrane is for hemodiafiltration. Ion exchange membrane is for Polymer Electrolyte Fuel Cell (PEFC), Biological Fuel Cells, ultrapure water, desalination, demineralization and so on.

There are two types exist for ion exchange membrane. One is cation exchange membrane and the other is anion exchange membrane.

Using Laplace equation $p_1 - p_2 = \frac{2\gamma_{iw}\cos\theta}{r}$ and Gibbs-Duhem equation $d\mu = -SdT + vdp$, the following equation can be obtained (Ishikiriyama, 1995).

$$\Delta T = \frac{2\gamma_{iw} v_w T_0 \cos\theta}{rL}$$

where $\Delta T$ is the depression of the melting temperature, $\gamma_{iw}$ is the surface energy between liquid water and solid ice, $v_m$ is the molar volume of water, $T_0$ is the melting temperature, $\theta$ is a contact angle, $r$ is pore radius, $L$ is the latent heat of fusion per mole, $\mu$ is the chemical potential, $S$ is the entropy, $p$ is the pressure. As shown in Fig. 2, the equation gives the results that $\Delta T$=45.4 K for pore size 1 nm and $\Delta T$=4.5 K for 10 nm.

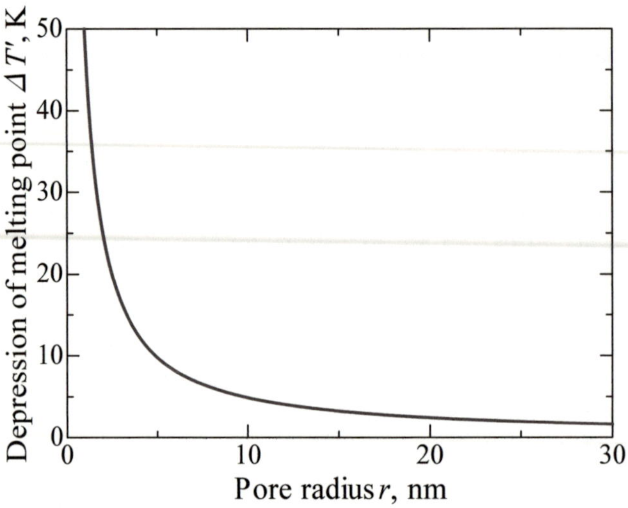


Fig. 2. Depression of the melting temperature due to a narrow space

Cation exchange membranes and anion exchange membranes with various thicknesses were used.

### 2.2 Porous elastic polymer membrane

The second membrane was made of porous elastic polymer material made by foaming. It is elastic and each pore is ideally independent to each other with cracks between them as shown in Fig. 3. So it is only semi-permeable when the membrane is expanded.

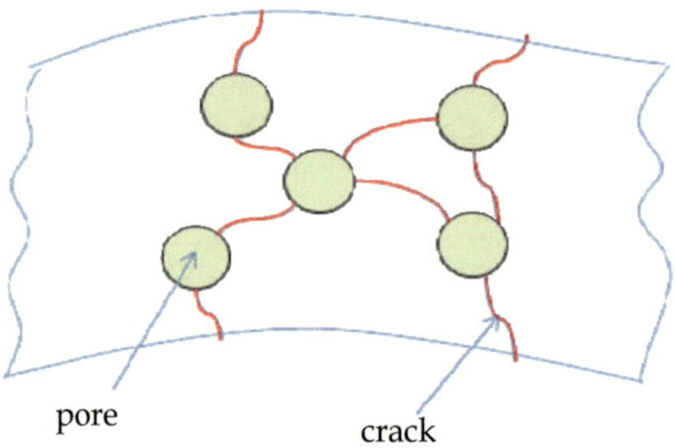


Fig. 3. Image of the cross section of porous polymer membrane

### 2.3 Styrene elastomer membrane

The third membrane was made of elastic polymer (thermoplastic elastomer) with a small hole in it. The material is chemical resistance to alcohol since a small amount of alcohol is in the contents of refrigerant. It has a high elongation and adequate tensile strength to keep water inside the capsule, especially at a low temperature. It is adhesive so the hole is closed when there is no volumetric expansion.

## 3. Propagation through ion exchange membrane

### 3.1 Experimental apparatus and experimental method

The type of ion exchange membrane tested in the research is shown in Table 2. Two types of cation exchange membranes and three types of anion exchange membranes were used. The purpose of the experiment was to check whether ice propagates through ion exchange membrane. The apparatus is shown in Fig. 4. The membrane was placed between two cells, namely the upper cell and the lower cell. Several kinds of concentration of NaCl solutions were prepared for both cells. The apparatus was kept under a constant temperature and artificially the nucleation was started in the upper cell under the supercooling state by installing the ice particle from the top of the needle as shown in Fig. 4. The ice gradually grew inside the needle and a single crystal ice appeared at the tip of the needle. Temperature of the solution in the upper cell was measured directly by inserting the thermocouple, and the temperature of the solution in the lower cell was measured indirectly from the outer surface of the cell to avoid natural nucleation due to the existence of the thermocouple. It was confirmed that there is almost no change in concentration during a series of experiments.

| Type | Cation exchange membrane | | Anion exchange membrane | | |
|---|---|---|---|---|---|
| name | CMT | CMV | AMT | AMV | DSV |
| thickness(μm) | 220 | 130 | 220 | 130 | 100 |
| Water content ratio | 0.28 | 0.27 | 0.21 | 0.22 | 0.29 |

Table 2. Type of ion exchange membrane

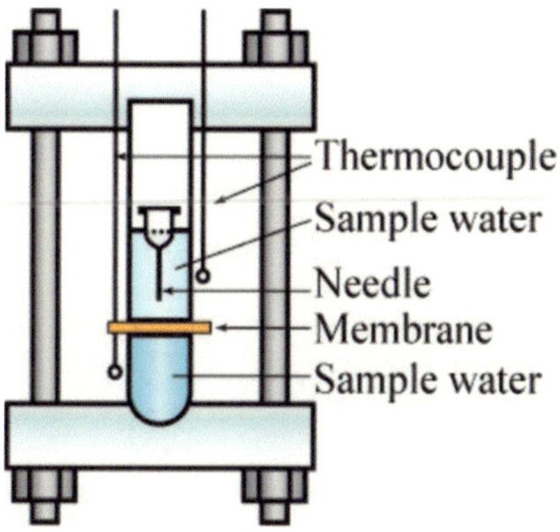


Fig. 4. Experimental apparatus

### 3.2 Results and discussions

Fig. 5 shows the results obtained. The abscissa shows the time after the ice touching the upper surface of the membrane. The ordinate shows the probability of propagation of ice to the lower cell. There were three degrees of supercooling tested, namely, $\Delta T$=3 K, 5 K and 7 K. As it can be observed from the figure, the only CMV membrane propagated ice. This is a cation exchange membrane with thin thickness. So, it can be said that a thin cation exchange membrane is better.

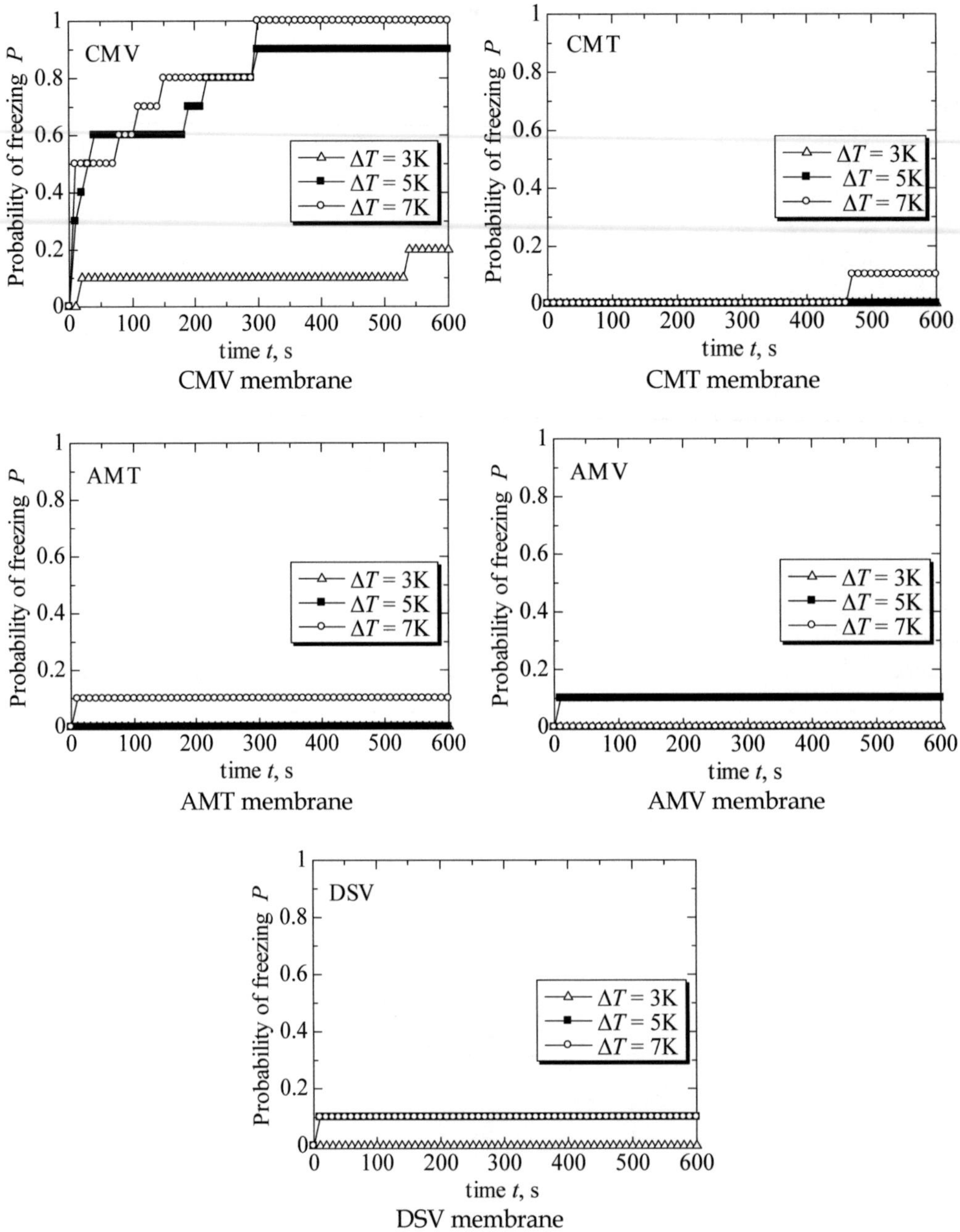


Fig. 5. Time wise variation of probability of propagation using various ion exchanger membranes

It may have a problem of durability.Since there is a volumetric expansion due to solidification, the capsule needs to have a function to absorb the volumetric change. Fig. 6 shows some of the example of having rubber material on the other side of the capsule. It has a demerit since the device becomes a little bit complicated.

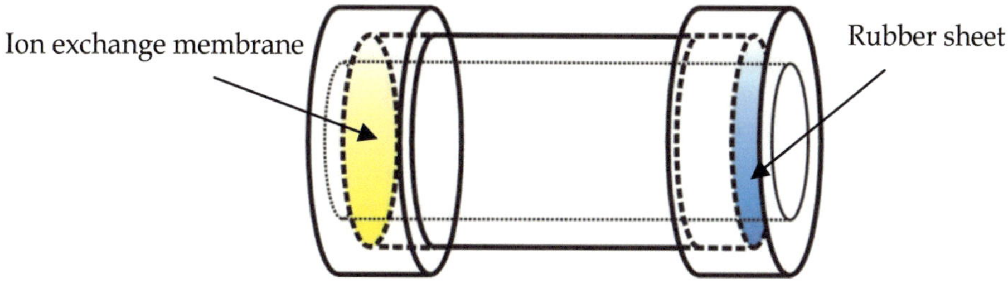


Fig. 6. Example of capsule having an ion exchange membrane on one side and a rubber sheet on the other side

## 4. Porous polymer material made by foaming

In order to solve the problem of complexity, an elastic porous membrane was selected. It is a porous polymer material made by foaming. Similar experiments to the one using the ion exchange membrane were carried out using the membrane. The results are shown in Figs. 7 & 8. Fig. 7 is for 10 wt% NaCl solution having the melting temperature of -6.56 °C and Fig. 8 is for 15 wt% NaCl solution having the melting temperature of -10.89 °C. It can be seen that in the case of 10 wt%, when water in the upper cell solidified, the temperature in the upper cell jumped to the melting temperature. However, since the solution in the lower cell was already in the supercooling state, ice propagated immediately with low degree of supercooling. In the case of 15 wt%, when water in the upper cell solidified, the temperature in the lower cell was above the melting temperature, so no propagation occurred at this stage. After the temperature reached down below the melting temperature for the solution, the ice propagated immediately with a low degree of supercooling.

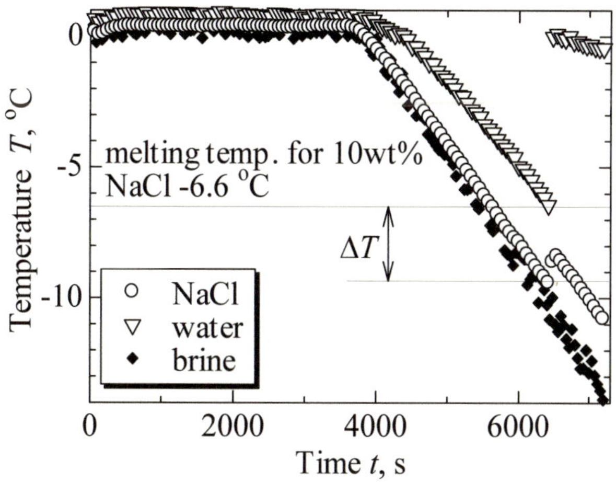


Fig. 7. Experimental result of propagation using porous polymer membrane with 10 wt% NaCl solution

Influence of the cooling rate on propagation of the ice was examined. Under two kinds of cooling conditions, temperature of the solution in the lower cell may differ to each other at the time of propagation. It is due to the existence of thermal resistance of the capsule. So the phenomenon was confirmed by the following method. The water in the upper cell was completely frozen and the whole apparatus was kept at the melting temperature of the solution in the lower cell. Then, experiments were started with constant cooling rates. The results are shown in Fig. 9. In both cases, there were ice propagations in 80 % of the experiments and the natural freezing in the lower cell occurred in 20 %. Fig. 9 shows that there is no difference in the degree of supercooling at propagation by changing the cooling rate. Hence it was found that the temperature of the membrane is the important factor for propagation of ice.

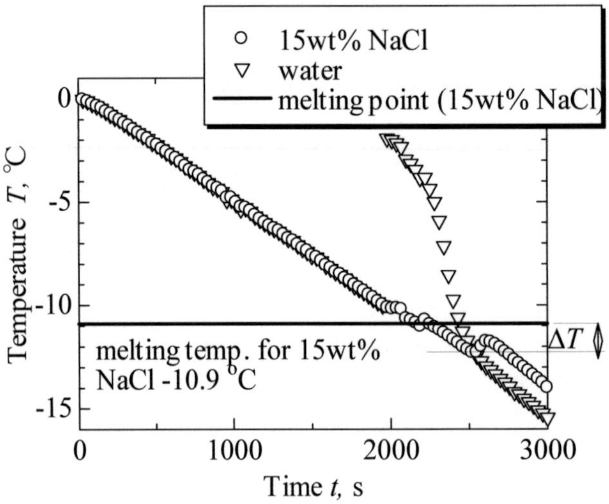


Fig. 8. Experimental result of propagation using porous polymer membrane with 15 wt% NaCl solution

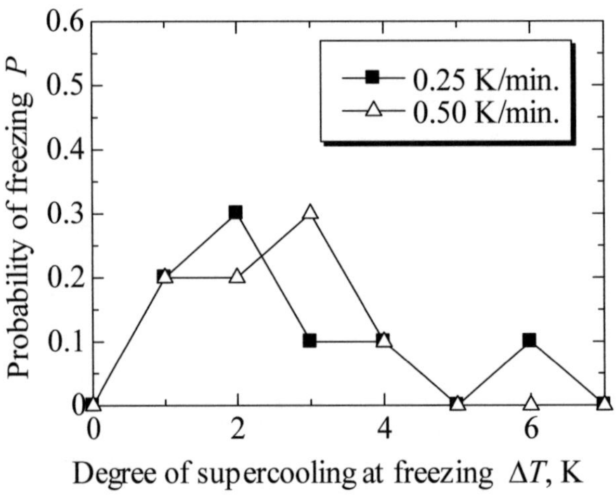


Fig. 9. Probability of propagation under two different cooling rates

By varying the location of ice appearance in the upper cell, it was found that the time taken for propagation of the ice and the probability of propagation are strongly influenced by the location of the ice appearance in the capsule. Especially when liquid in the upper cell was a solution, the probability of propagation became low when the ice appeared at the location far away from the membrane. The reason seems to be that the temperature in the upper cell rises above the melting temperature for the lower cell and also concentration in the upper cell near the membrane becomes higher due to the elimination of solute from the ice during the solidification. Hence the melting temperature in the upper cell becomes lower which leads to a decrease of the degree of supercooling in the upper cell.

Capsules were made using porous polymer membrane. The material can easily be melted by heating, so as shown in Fig. 10, one sheet of membrane was put on top of the other having a spacing material between them. The reason for the spacing material to put between them was to avoid water in the capsule to be vanished away from the capsule due to the osmotic pressure. The circumference was sealed by heating. Air in the capsule was removed from the pipe and water was installed instead. The thermocouple was inserted from the pipe as well. The diameter of the membrane was 25 mm.

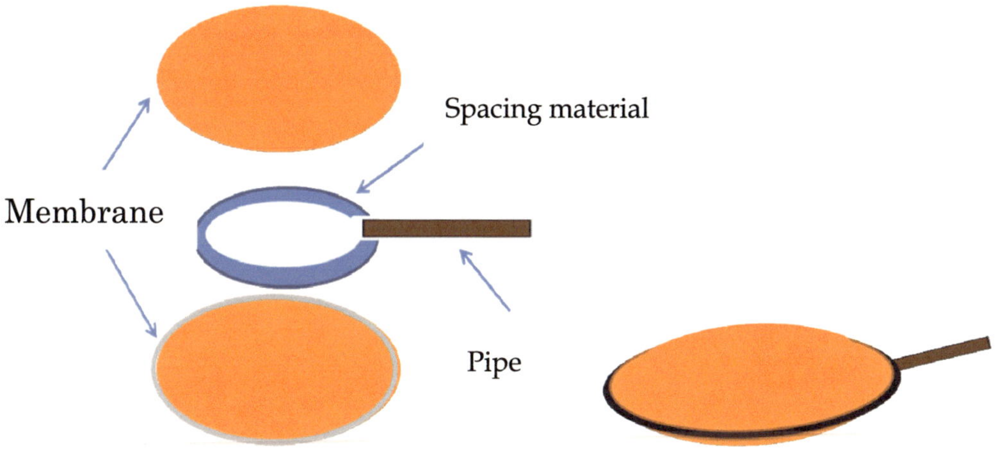


Fig. 10. Capsule made of porous polymer membrane

100 ml of 15 wt% of NaCl solution was put inside the beaker and the capsule was installed in the solution. The beaker was cooled down with a constant cooling rate and the temperature at ice propagation was measured. Two types of membranes were selected. One membrane had a thickness of 2.2 mm, the average pore size of 7 μm and the porosity of 66 %. The other membrane had a thickness of 3 mm, the average pore size of 7 μm and porosity of 75 %.

The results are shown in Fig. 11. The figure shows the frequency distribution against the degree of supercooling at propagation. It can be seen that in both cases, propagation occurred at around 1 K of the degree of supercooling. So it can be said that the capsule is effective.

The instant of propagation was observed. The typical example of the propagation was shown in Fig. 12. The camera was set at the side of the capsule in order to observe the ice appearance on the membrane. It can be seen that ice grew slowly from the membrane.

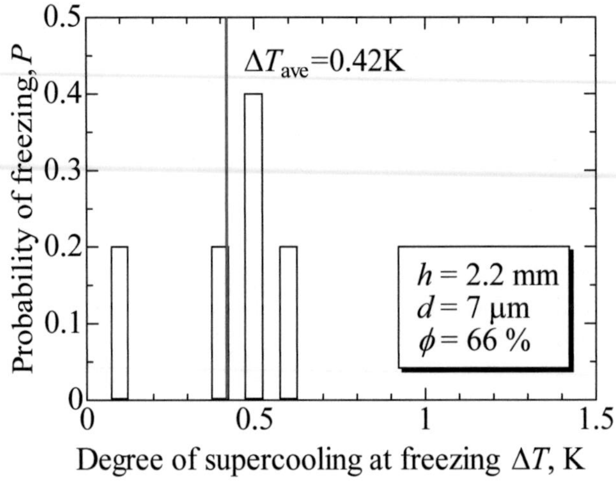


(a)

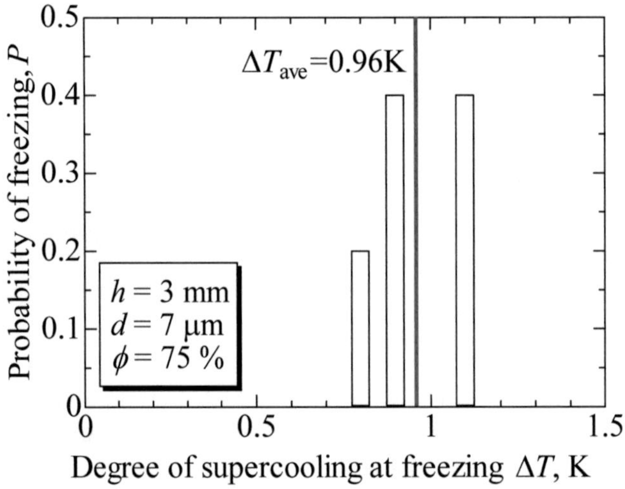


(b)

Fig. 11. Frequency distribution of ice propagation through porous polymer membrane

Fig. 13 shows a typical example of the difference with and without using the capsule. It can be seen that the installation of the capsule was effective for propagation of ice. Fig. 14 shows a typical example of the performance under the repeated use of the capsule. It can be seen that the capsule is effective for a repeated use. However, by repeating the experiment and measuring the concentration inside the capsule, it was found that there was an individuality in the quality of the membrane, such as pore size and the distribution of the pore in the membrane. So, it was rather difficult to produce a membrane with a constant quality.

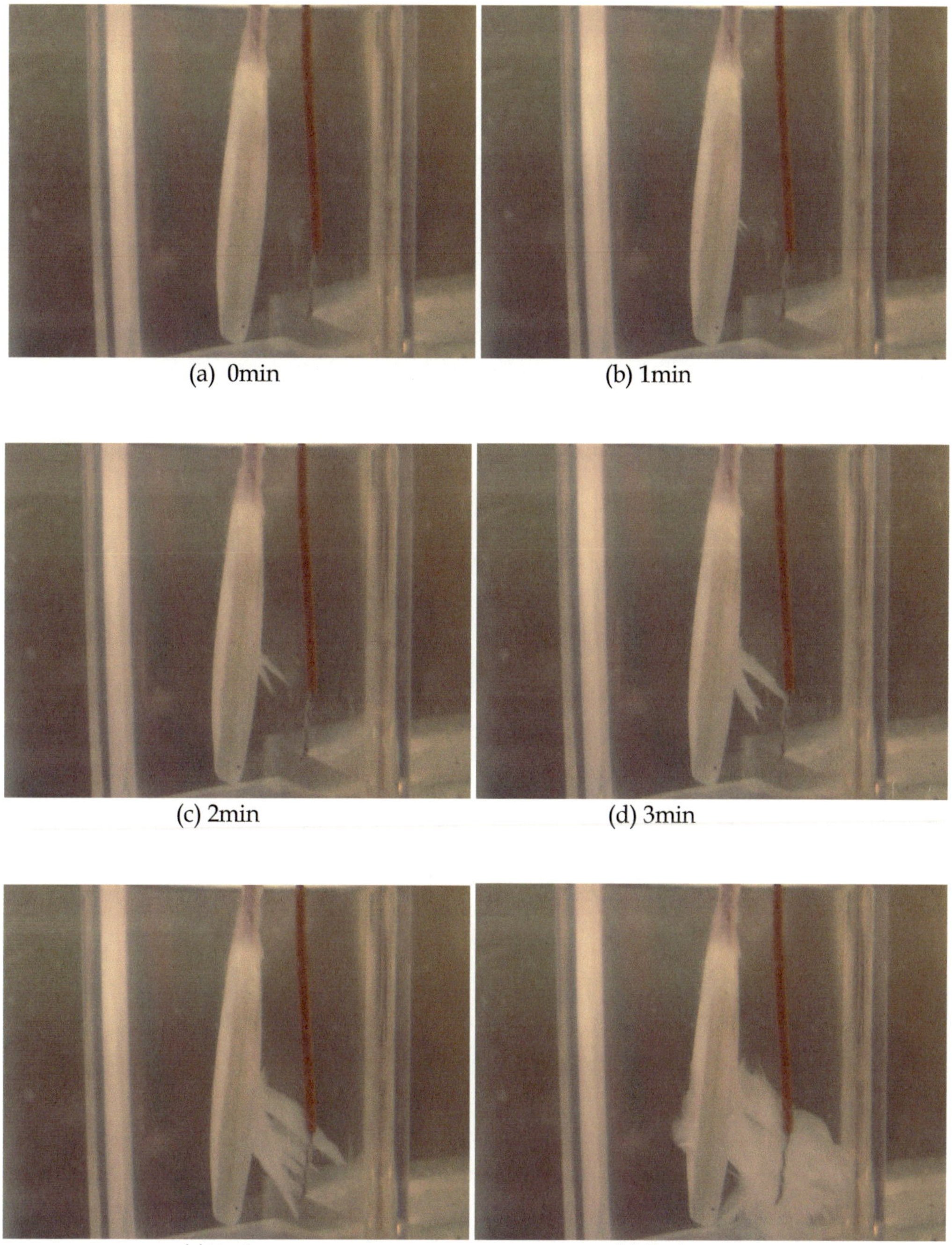

(a) 0min (b) 1min

(c) 2min (d) 3min

(e) 4min (e) 5min

Fig. 12. Typical example of photos during propagation

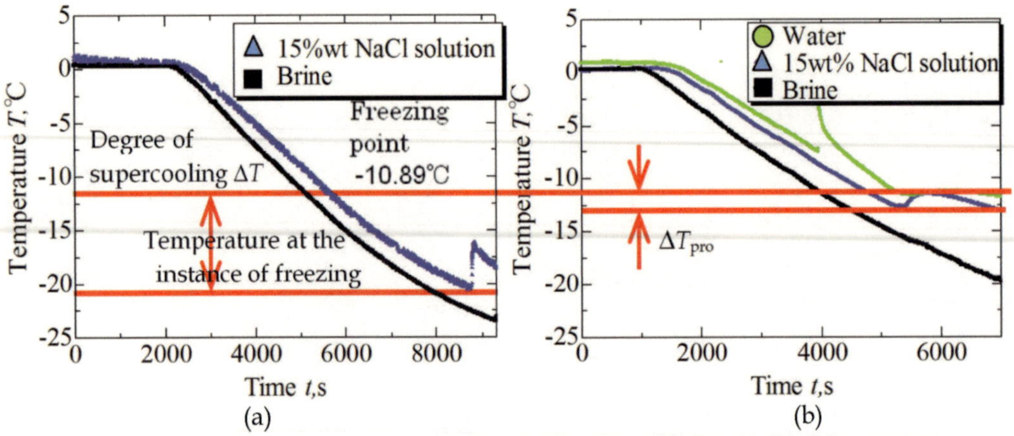


Fig. 13. Typical example of freezing of solution with and without capsule (a) without capsule, (b) with capsule

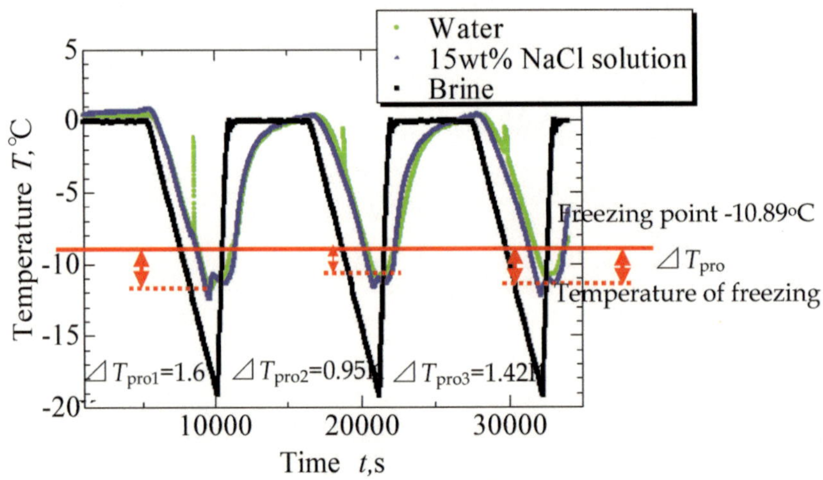


Fig. 14. Typical example of the performance under the repeated use of the capsule

## 5. Styrene elastomer membrane with pinhole

The details of the capsule are shown in Fig. 15. The main body of the capsule is made of polypropylene tube. The open part of the two tubes was facing to each other having 2 mm gap between them. Then styrene elastomer membrane with several pinholes was rolled to cover the gap. The membrane can easily be melted by heating, so it can be glued together. The styrene elastomer has a phenomenon that it is elastic at a low temperature. So, the membrane was enlarged in one direction up to 300 % and 6 holes having 0.4 mm diameter were created. It was confirmed that the hole was closed under the condition of no enlargement, and the elasticity was kept at the low temperature.

The logic of the propagation of freezing using the capsule is explained as follows. There is a difference in melting temperature between the refrigerant in a cool box and water in the capsule. After installing the capsule in the cool box, the cool box is cooled down. Then the water in the capsule starts to freeze first because the melting temperature is higher. When the temperature in the cool box reaches the melting temperature of refrigerant, the water in the capsule is frozen completely. The volume of the capsule is expanded due to phase change of the water contained inside. So, the membrane is expanded and ice in the capsule is uncovered to the refrigerant through pinhole. Then the pinhole becomes a trigger for the refrigerant to freeze with a low degree of supercooling.

Experiments were carried out by preparing 100 ml of 15 wt% of NaCl solution in a beaker. The capsule was also installed inside the solution. Temperature inside the capsule and the solution near the capsule were measured. The melting temperature of the solution was -10.89 °C. The solution in the beaker was cooled down with a constant cooling rate of 0.15

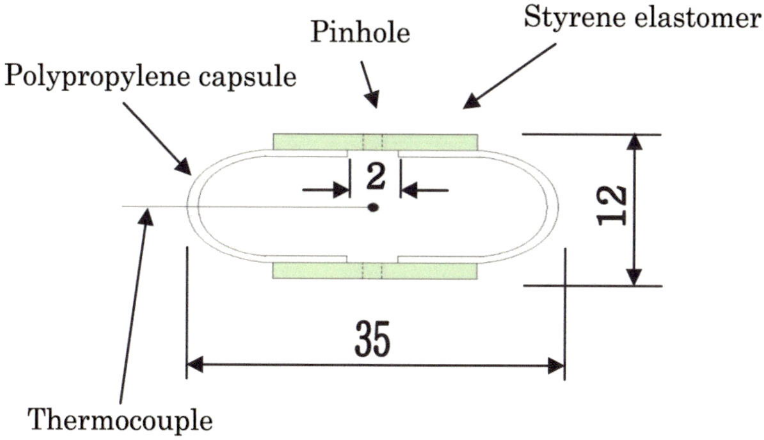


Fig. 15. Capsule using styrene elastomer

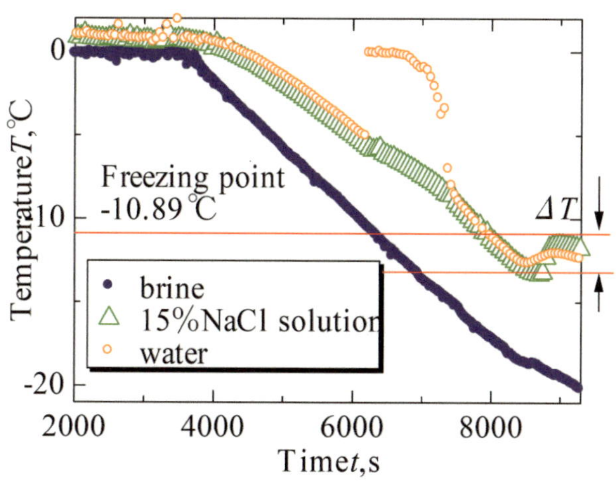


Fig. 16. Typical example of temperature record in a case of 15 wt% NaCl with a styrene elastomer capsule installed

K/min. The temperatures in the capsule and outside the capsule were measured and the propagation of freezing was checked. Due to the freezing of the water inside the capsule, the middle part of the capsule expanded and the size of the pinhole became bigger.

(a) 0 s

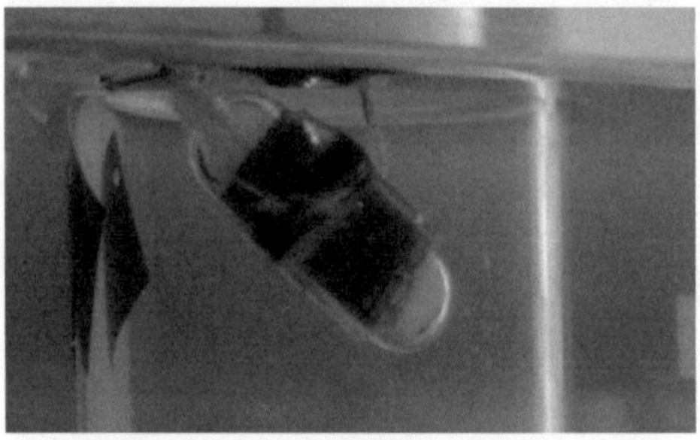

(b) 30 s

(c) 60 s

(d) 180 s

Fig. 17. Typical example of ice propagation using styrene elastomer capsule

One of the typical examples of the time wise variation of the temperature is shown in Fig. 16. Due to the solidification of water inside the capsule, the temperature of the water jumped to 0 °C. Then, the water kept the temperature for a while and it started to drop down again until the temperature of the solution started to jump. It was the instance of propagation of freezing. The degree of supercooling at freezing was around 2 K.

One of the examples of propagation view is shown in Fig. 17. Since the water inside the capsule was frozen completely at this moment, the color inside the capsule looked white and the middle part of the capsule swelled. The original color of the membrane was transparent and it was difficult to observe the propagation on the surface. So, the inner surface of the membrane was painted in black. From the photos it can be observed that the ice propagation started at two points on the membrane and the ice grew slowly. The reason for the slow growth was because of the low degree of supercooling and the high concentration of the solution.

## 6. Conclusion

A method to induce solidification of supercooled refrigerant using membranes was introduced. Three kinds of membranes were selected, namely, ion exchange membrane, porous elastic polymer membrane and styrene elastomer membrane. The following conclusions were made.

1. Cation exchange membrane and anion exchange membrane with different thickness were examined. It was found that cation exchange membrane with thin thickness propagated the ice.
2. Porous elastic polymer membrane was used to establish the capsule and experiments were carried out. It was found that the membrane was effective for propagating the ice with a low degree of supercooling. Influence of various factors on propagation of ice through membrane, such as difference in the melting temperature between the refrigerant and the liquid material, difference in the cooling rate, difference in the location of the initial ice appearance, difference in the thickness of membrane, were investigated. It was confirmed that the refrigerant with a low melting temperature can be frozen with a small degree of supercooling by inserting a capsule with water or solution having a higher melting temperature in it and a sheet of membrane to separate between them. The propagation time becomes longer when the difference in the melting temperature of two solutions is big. It is because the solution with higher melting temperature solidifies before the solution with the lower melting temperature reaches its solidification temperature. The temperature rises due to solidification and it needs time to lowdown the temperature again.
3. Styrene elastomer membrane which has high elasticity at low temperature was selected and a small pinhole was put through it under the expanding condition in one direction. The shape of the capsule to suit with the characteristics of expansion was prepared and water was put in the capsule. The capsule was set in the refrigerant and the propagation of freezing with a low degree of supercooling was investigated, experimentally. As a result, due to the volume expansion by freezing, the pinhole on the membrane was expanded and it became a trigger for the refrigerant to freeze under the degree of supercooling at around 2 K. Hence, it can be proved that the idea of installing the capsule can suppress the supercooling phenomenon.

## 7. References

Hozumi T., Saito A. & Okawa S. (1999). Research on Effect of Ultrasonic Waves on Freezing of Supercooled Water, *6th International Symposium on Thermal Engineering and Science for Cold Regions*, 65-72

Hozumi T., Saito A., Okawa S. & Matsui T. (2002). Freezing Phenomena of Supercooled Water under Impacts of Ultrasonic Wave, *International Journal of Refrigeration*, 25, 7, 948-953

Hozumi T., Saito A., Okawa S. & Matsumura T. (2002). Effect of Bubble Nuclei on Freezing of Supercooled Water, *International Journal of Refrigeration*, 25, 2, 243-249

Hozumi T., Saito A., Okawa S. & Eshita Y. (2005). Effects of shapes of electrodes on freezing of supercooled water in electric freeze control, *International Journal of Refrigeration*, 28, 389-395

Inada T., Zhang X., Yabe A. & Kozawa Y. (2001). Active control of phase change from supercooled water to ice by ultrasonic vibration, 1. Control of phase-change temperature, *International Journal of Heat and Mass Transfer*, 44, 4523-4531

Ishikiriyama K., Todoki M. & Motomura K. (1995). Pore size distribution (PSD) measurements of Silica gels by means of differential scanning calorimetry, *Journal of Colloid and Interface Science*, 171, 92-102

JSME Date Book Thermophysical Properties of Fluids (1983). 468 (in Japanese)

Okawa S., Saito A. & T. Harada T. (1997). Experimental Study on the Effect of the Electric Field on the Freezing of Supercooled Water, *Proc. Int. Conf. on Fluid and Thermal Energy Conversion '97*, 347-352

Okawa S., Saito A. & Fukao T. (1999). Freezing of Supercooled Water by Applying the Electric Charge, *5th ASME-JSME Thermal Engineering Joint Conference*, (in CD-ROM)

Okawa S., Saito A. & Minami R. (2001). The Solidification Phenomenon of the Supercooled Water Containing Solid Particles, *International Journal of Refrigeration*, 24, 1, 108-117

Okawa S., Saito A., Kadoma Y. & Kumano H. (2010). Study on a method to induce freezing of supercooled solution using a membrane, *International Journal of Refrigeration*, 33, 7, 1459-1464

Okawa S. & Taniguchi Y. (2010). Fundamental research on freezing of refrigerant by making use of a membrane, *The 5th Asian Conference on Refrigeration and Air-conditioning*, 042, (in CD-ROM), (Tokyo, Japan, June 2010).

Saito A., Okawa S., Tojiki A., Une H. & Tanogashira K. (1992). Fundamental Research on External Factors Affecting the Freezing of Supercooled Water, *International Journal of Heat and Mass Transfer*, 35, [10] 2527-2536

Shichiri T. & Araki Y. (1986). Nucleation Mechanism of Ice Crystals under Electrical Effect, *J. Crystal Growth*, Vol.78, pp.502-508

5

# Supercooling at Melt Growth of Single and Mixed Fluorite Compounds: Criteria of Interface Stability

Jordan T. Mouchovski
*Institute of Mineralogy and Crystallography*
*Bulgarian Academy of Sciences*
*Bulgaria*

## 1. Introduction

### 1.1 Single and mixed fluoride crystal compositions – Properties, peculiarities and application

The artificially grown single fluoride crystal compositions of alkali-earth metals ($MF_2$, $M$ = $Ca, Sr, Ba$) as well as the congruently melted compounds of the type $Ca_{1-x}Sr_xF_2$ or $Ca_{1-y}Ba_yF_2$ possess a high cubic crystal symmetry structure that determines isotropic optical properties of these materials and correspondingly - their broad usability in various optics devices. Amongst the single fluoride, $CaF_2$ stands out by its unique optical, physical, and mechanical characteristics: light transmission from far ultraviolet (UV) up to middle infrared (MIR), impressive dispersive power and partial dispersion, controllably low stress-induced birefringence, negligible solubility in water and insolubility in many acids, adequate hardness, fairly good thermal conductivity, high resistance to radiation at low luminescence level. The ability of this crystal being a good matrix for rare earth (RE) and actinide completes its superior characteristics and determines exclusively wide application - by variety of optical elements - in UV/MIR and laser optics, holography, dosimetry, lithography, materials processing and other.

Similarly to $CaF_2$, the matrix of several congruently melted alkali earth (AE) fluoride crystal systems of the type $M_{1-x}M'_xF_2$ ($M \neq M'$) can contain highly concentrated activator RE ions. The physical-chemical properties of such systems, being put under effective control by adjustment of growing conditions, provide production of crystals with desired characteristics. To gain experience in this direction is as much important as the applicability of the mixed AE fluoride compounds expands as: windows in powerful optical quantum generators for laser thermonuclear synthesis, excellent luminescence coverage, Anti-Stokes luminophors, radiation dosimeters, various devices for preservation and optical processing of information and others.

Other mixed systems, formed by congruent melting and crystallization of AE and RE fluorides $MF_2$–$REF_3$, represent a high-potential source of new materials in many fields of science and technology. Peculiarity of such systems is that the introduction in the matrix of large proportion of RE ions may result in formation of hetero-valence, strongly non-stoichiometric solid solutions of the type $M_{1-x}RE_xF_{2+x}$, Here $x \leq 0.5$ and such imposed variations in solid solution composition can be controlled precisely so that to ensure structural micro-homogeneity of the grown crystals with, eventually, some extraordinary physicochemical properties: mechanical (increasing considerably micro-hardness without any cleavage), electrical (amplifying ion conductivity by several orders of magnitude) as well as spectroscopic (strongly improved optical parameters). These type solid solution fluoride crystal systems are used as: structural materials in UV–IR optics; solid electrolytes with high fluorine ion conductivity fast radiation-resistant scintillators, photo-refractive materials, substrates with controlled lattice parameter and low-temperature heat-insulating material; in quantum electronics as host materials and as active elements in solid-state lasers.

### 1.2 Growing techniques – Interface stability at melt growth

*Melt growth* of any single or mixed fluoride systems possesses substantial advantages over *growth from solutions* because it may provide conditions for growing homogeneous crystals with two–three orders of magnitude higher crystallization rate (CR). Understandable, the melt growth turned out to be the only method that is appropriate for developing effective techniques for production of crystals with high optical quality (Yushkin, 1983). Specifically, various modifications of Bridgman-Stockbarger (BS) technique are usually employed whereat a container with molten material is moved in vertical two-zone (or multi-zone) furnace, the axial *T*-gradient of which provides a gradual melt/crystal phase transition (Mouchovski, 2007a). Under thermodynamically grounded segregation effect on liquid/solid interface ($IF_{L/S}$), the distribution of components (impurities) in growing crystal body may become inhomogeneous that causes relevant non-uniformity in its optical characteristics. Nevertheless, in many cases when growing mixed fluoride systems, the residual cationic impurities are of trace concentration so that their impact upon general components' distribution is left negligible.

The growth kinetics itself differs depending on the heat and entropy of melting for particular system, thus determining either face-type (atomically smooth surfaces) or normal (atomically rough surfaces) growth mechanism. As being proved for crystals with fluorite type structure, their heat and entropy of melting turn out sufficiently low for melt growing to proceed by "normal growth" mechanism according to Jackson criterion (Chernov, 1984; Jackson, 2004). In this case the fluoride admixture enriches the crystallization zone (CZ), owing to which $IF_{L/S}$ loses stability in conformity with the degree of so called *constitutional melt supercooling* (CMSc).

## 2. Melt supercooling as physical-chemical phenomenon

Supercooling phenomenon represents a thermodynamic process of lowering the temperature $T$ of a liquid or gaseous substance below its freezing point $T_{fr}$ without solid phase transition to occur. Below $T_{fr}$ crystallization can start in any liquid (melt) in the presence of either spontaneously arisen nucleus or artificially introduced seed crystal or nucleus. When lacking any such nucleus the liquid may stay down to the temperature whereat homogeneous nucleation to start.

Supercooling arises inevitably when artificial crystals have been grown by BS method, that is, in a container moving downwards in parallel to gravity force temperature field with negative $T$-gradient. The phenomenon appears within a very short zone (layer) ahead the crystallization front (CF) as a result of constitutional changes around the moving $IF_{L/S}$ or/and specific heat transport throughout the loaded container (crucible) possessing a complex thermal conductivity. No matter what mechanism results the supercooling, it may affect the smoothness and stability of the CF in micro-level, this way accelerating spontaneous growth of protuberating structures on the $IF_{L/S}$ resulting in growth of crystals with cellular structure.

The nature of CMSc is liable for explanation on the grounds of *conventional crystallization theory* according to which, an extremely fast ionic exchange provides a local quasi-equilibrium on the $IF_{L/S}$. This equilibrium can be disrupted owing to inconformity between the CR and the effective diffusion rates of the ions throughout the supercooled melt zone before their incorporation into the structure of growing crystal. This way the crystal-chemical reactions nearby the $IF_{L/S}$ are diffusion-controlled that is manifested on the phase diagram as difference between liquidus and solidus curves. Under such conditions, the specific change in melt composition ahead the $IF_{L/S}$ causes its geometric and/or morphological instability: the polyhedral crystal form stays unstable, sprouting protrusions at its corners and edges where the degree of supersaturation attains the highest level. The process leads to accelerating spontaneous growth of protruding structures on the $IF_{L/S}$ and growth of crystal with cellular substructure, which is transformed into a dendroid structure.

## 3. Criteria for interface stability

### 3.1 Constitutional supercooling principle

For concentrated, two metal component, congruently or incongruently melted, representing mixed fluoride systems of the types $MF_2$—$M'F_2$ and $MF_2$—$REF_3$, the simplest one-dimensional model for $IF_{M/Cr}$-instability considers a normal mechanism in Tiller-Chalmers microscopic approximation (Tiller, 1953), where several factors - the released heat and density jump at crystallization, the contribution of surface energy as well as the anisotropy of segregation coefficient are not taken into consideration. Besides, the low heat of melting for these fluoride systems implies the heat-transport to be neglected at the account of a vast mass transport. Nevertheless when growing large sizes crystals, the mass transport may be concurred strongly by a heat-transport owing to significant radiation flow throughout the growing transparent crystal, surpassing increasingly the conductive flow therein (Deshko et al., 1986, 1990). Disregarding this case, if the thermal field inside the furnace may ensure $T$-gradient in the melt at the CF, $G$, that is larger than the gradient of the equilibrium solidification temperature $T_L$, the derivation of criterion for $IF_{L/S}$-stability starts from the inequality (Kuznetsov & Fedorov, 2008).

$$G = (dT/dz)_{CF} > dT_L/dl \qquad (1)$$

where $l$ is the thickness of the short layer zone ahead the CF.

Further, the material balance at liquid-solid interface is considered relating the CR $V_{cryst}$, that is, the growing rate, to interdiffusion coefficient (solute diffusivity) $D$ by the equation:

$$V_{cryst}(x_S - x_L) = -D(dx/dl) \qquad (2)$$

Here $x_S$ and $x_L$ are the molar concentrations of solid and liquid phases respectively and $(dx/dl)$ – the concentration gradient at the CF.

Eq. (2) unifies the two types mass transport, the jump-like forcing-out (when the equilibrium coefficient of segregation $k_o < 1$) and absorption of the second component and its diffusion into the melt bulk or from the bulk (when $k_o > 1$).

Using the mathematical expression:

$$dT_L/dl = (dT_L/dx)(dx/dl) = m(dx/dl) \tag{3}$$

where the slope $m$ of the liquidus curve (tangent of the slope's angle for the liquidus curve) is being introduced, from (1) and (2) it follows:

$$G\,D/V_{cryst} > m\,\Delta x \tag{4}$$

Here $\Delta x = x_S - x_L$ expresses the apparent discontinuity in component's concentration at the CF.

The inequality (4) is an extension to region of the high concentrations and non-stationary processes of Tiller–Chalmers stability criterion (Chalmers, 1968):

$$G/V_{cryst} > -\,m(1 - k_o)x_{o(b)}/k_o\,D \tag{5}$$

where: $x_{o(b)}$ is the admixture molar concentration of the second (impurity) component in the melt prior to solidification, equal to its mean concentration in the melt bulk (sufficiently far from the CF), while $k_o$ is defined by the ratio of solid $x_{oS}$ to liquid $x_{oL}$ phase concentrations for this component, specified at the CF.

This criterion, derived under suggestions of stationary crystallization caused by CMSc, small concentration for the second component and constant values for $k_o$ and $m$, has been proved being valid for concentrated melts as well (Kaminskii et al., 2008; Sobolev et al., 2003).

The solution of the van't Hoff equation:

$$d\ln k_o/dT = L_{molt}/RT^2 \tag{6}$$

for the case of coexisting phases in binary systems with low content of the second component gives the expression:

$$m = \Delta T/\Delta x_o = (RT^2/L_{molt})(k_o - 1) \tag{7}$$

where $R$ is the universal gas constant and $L_{molt}$ is the enthalpy of melting (latent heat of fusion) for matrix material (the first component).

The application of later equation makes it possible to determine $m$ from the liquidus curve at $x \to 0$ (initial slope of the curve, $m_{init}$) and, correspondingly, $k_{olim}$ – the limiting segregation coefficient of the admixture when its concentration tends to zero. Thence:

$$k_{olim} = (L_{molt}/RT^2)m_{init} + 1 \tag{8}$$

Substituting (7) for $m$ in (5) the criterion for interface stability takes the form:

$$G\,D/V_{cryst} > -(RT^2/L_{molt})(k_o - 1)(1 - k_o)x_{o(b)}/k_o = (RT^2/L_{molt})x_{o(b)}(1 - k_o)^2/k_o \quad (9)$$

which, in case of slightly soluble into the lattice impurities ($k_o$<< 1), can be approximate by:

$$G\,D/V_{cryst} > (RT^2/L_{molt})x_{o(b)}[(1/k_o) - 1] \quad (10)$$

The physical explanation of criterion (10) is grounded on so called *segregation mechanism*, under which impact are all kind of impurities penetrated in trace concentrations within the layer ahead the moving CF. These impurities are usually several metal cations and oxygen containing anions, any concentration gradients of which arisen can not be equalized at a lack of sufficiently intensive convection into the melt bulk and relatively low diffusion rates. Hence, kinetic limitations determine the real crystallization conditions liable for impurities' segregation at the CF. Mathematical this phenomenon is convenient being expressed by introducing in (5) an extra multiplayer, $k_{eff}$, the so called effective coefficient of segregation (Sekerka, 1968):

$$G/V_{cryst} > -(m\,x_o/D)[(1 - k_o)/k_o.]k_{eff} \quad (11)$$

Here:

$$k_{eff} = k_o/[k_o.+ (1- k_o)exp(-\Delta)] \quad (12)$$

where the index $\Delta = V_{cryst}\,\delta/D$ has a meaning of "*reduced CR*" and $\delta$ is the thickness of the diffusion boundary layer.

Evidently the parameter $\Delta$ decreases with lowering $V_{cryst}$ or $\delta$ and with an increase of $D$ whereat the denominator in (12) enhances, tending to 1 at $\Delta \rightarrow 0$. Correspondingly $k_{eff}$ goes down, tending to the equilibrium segregation coefficient $k_o$. The other limiting case ($\Delta \rightarrow \infty$) leads to $k_o \rightarrow 1$, that is, the impurity's content in the crystal stays equal to that in the melt bulk.

The position of $k_{eff}$ as regards $k_o$ depends on whether the equilibrium coefficient stays lower or higher than unity: if $k_o > 1$, then the inequalities $1 < k_{eff} < k_o$ are being fulfilled. Opposite to that, if $k_o < 1$ (as in the cases for mostly RE in fluorite matrix (Nassau, 1961), then $k_o < k_{eff} < 1$.

The thickness of the diffusion boundary layer $\delta$, assessed approximately of the order of 0.1 cm, can be substantially reduced in case of stirring the melt whereat it approaches a limiting value of 0.001 cm, conditioned by liquid to solid adhesion. Thus, providing sufficiently intensive stirring of the melt, one may anticipate considerable enhancement of $IF_{L/S}$-stability. Without such stirring, owing to low diffusion rate in liquid phase for the heterogeneous impurities, they either enrich (at $k_{eff} < 1$) or impoverish (at $k_{eff} > 1$) the layer ahead the $IF_{L/S}$. This way attained enhancement/lowering of impurities' concentration causes a relevant local lowering of melt temperature in the layer, followed by rejection of the impurities from crystallizing surface and their embedment in crystal lattice. At that the equilibrium freezing temperature in the melt adjacent to CF remains above the real temperature established. As a consequence, numbers of nucleolus centres appear simultaneously in supercooled layer that breaks the morphological $IF_{L/S}$ stability. In conformity with that the normal growth turns out replaced by high rate poly-crystalline (dendroidal) solidification with formation of cellular sub-structure. Such trapped impurities appear as well light-scattering centres lowering the light transmission within corresponding spectral range.

The approximation (10), giving deviation below 1% for $k_o \leq 0.01$, is especially convenient to demonstrate clearly how it could be attained specific conditions, guaranteeing the interface stability, varying appropriately definite substantial and/or apparatus factors so that the influence of the second (impurity) component to be neglected.

### 3.2 Function of stability and interdiffusion coefficient

The right-hand side of inequality (4) is denoted by $F(x)$ as function of stability (P.I. Fedorov & P.P. Fedorov, 1982; Turkina et al, 1986; Van-der-Vaal's & Konstamm, 1936), possessing dimensionality of temperature. The function is nonnegative $F(x) \geq 0$, vanishing at points with pure components and at the extremes (points of congruent melting of the solid solutions) where the liquidus and solidus have gotten tangential points. In the range of small impurity's concentrations $F(x)$ approximates a straight line (Djurinskij & Bandurkin, 1979). To a first approximation $F(x)$ can be expressed by the $T$-difference between liquidus and solidus curves.

The function of stability reveals definite physical meaning as regards the real crystallization process: there is no concentration supercooling and the CF should keep a planar and stable form for any cases when the value of parametric combination $G\,D/V_{cryst}$ lies higher at a given $x_s$ than the corresponding point pertaining to $F(x)$-curve. In this combination the furnace design and power supply distribution are decisive for determining the steepness for G. After establishment of this gradient, $V_{cryst}$ is under control either by altering the speed of crucible (SC) withdrawal or by cooling slowly the immobile load by Stober growing technique so that both cases the growth to proceed within constitutional regions of interface stability. Such approach requires an eventual compositional dependence for interdiffusion coefficient $D$ to be searched by using an expression, following from (4):

$$D = m\,\Delta x(V_{cryst}/G) = F(x)_{crit}(V_{cryst}/G) \tag{13}$$

where $F(x)_{crit}$ is the critical stability function upon the transition from stable plane CF to unstable one, the values of which function are calculated with sufficient precision from corresponding compositional phase diagram for particular fluoride system. Here special attention should be paid to number of points, for which dataset precision is *in situ* high: the melting temperatures for the end members of studied solid solution, the extremes (maximums or minimums) on the melting curve, and eutectic equilibriums.

### 3.3 Perturbation theory of interface stability

The inequalities (5), respectively (9), appear appropriate criterion for quantitative analysis about the influence of any impurities on melt supercooling arising ahead the CF in accordance with their amount and solubility into the structure of growing crystal. Processing the experimental data on hand, concerning $IF_{L/S}$ stability of many single crystal compositions and metal alloys, Sekerka (1968) outlined mostly of them being in satisfactory agreement with inequalities (5) or (9). However abiding this criterion, based on purely thermodynamic considerations, failed in several other cases due to its shortcomings. Indeed, the constitutional supercooling principle refers only to liquid phase, thus ignoring the latent heat of fusion and the heat flow throughout the growing crystal. Unlike it, the *perturbation theory of interface stability* has been developed being based upon the dynamics of the whole melt-crystal system (Mullins & Sekerka, 1963, 1964; Sekerka, 1965, 1967, as cited in Sekerka,

1968). Besides, this theory provides a description of time evolution of the perturbed interface, following the alterations in thermal and concentration fields. It seams this theory should replace that of CMSc principle despite giving a much more complicated expression for perturbation $IF_{L/S}$ stability criterion (Sekerka, 1965). One may assess the reasonability of such replacement after writing the perturbation $IF_{L/S}$ stability criterion in a form usable for comparative analysis to inequality (5):

$$(G/V_{cryst}) + (L_{molt}/2K_L) > [-m(1-k_o)x_{o(b)}/k_o D][(K_S + K_L)/2K_L]\,\varphi \quad (14)$$

Here, the latent heat of fusion $L_{molt}$ refers to unit volume of the solid crystal phase; $K_L$ and $K_S$ present the thermal conductivities of the liquid and solid phase, respectively; $\varphi$ is function of stability that depends in a complex way on $k_o$, $m$, $x_{o(b)}$, $D$, $V_{cryst}$, the absolute melting point of the pure solvent (first component), and $\Gamma$-function, that expresses melt-crystal surface free energy.

Since $\Gamma \leq 0.001$, $\varphi$ is usually between 0.8 and 0.9, approaching 1 for negligibly small surface free energy. Such relatively small deviations of $\varphi$ from unity illustrate a tendency of the surface free energy to stabilize slightly the $IF_{L/S}$. Here some comments are useful: the divergence between the results, obtained by criterions (5) and (14), becomes greatest for small concentrations, $x_{o(b)}$, of the second (impurities') component. Thence, for the first, highly concentrated component (the matrix), it exists a certain limiting concentration for the second component below which, even at very fast crystallization rates, a cellular structure could not be formed so that the established planar CF would be absolutely stable. Nevertheless, with increasing the concentration of the second component, the contribution of the surface energy lessens rapidly and may be neglected. It should be notice as well that it is possible the corrections due to surface energy stabilizing effect to become significant around the extreme points upon the melting curve for several mixed fluoride solid solutions.

The ratio $[(K_S + K_L)/2K_L]$ in (14) may affect tangibly the $IF_{L/S}$-stability only if $K_S$ is appreciably higher than $K_L$. Opposite to that, when thermal conductivities of the two phases are close to each other, this factor stays insignificant for assisting the criterion (14) being fulfilled. However in case the growing medium is optically transparent (as appear mostly single and/or mixed fluoride compounds), not conductivity but radiation should determine the heat transport throughout the load. Besides, the melts of these compounds turn out semi-transparent. Despite the thermal processes for such liquid-solid systems have been described by a system of integral-differential equations solved at specific boundary conditions, the developed on this grounds non-linier stability theory leads inevitably to considerable mathematical complications and difficulties for correct interpretation of the results. Both later could be overcome by using a simple model where thermal conductivities were replaced by their effective analogs, representing a sum of thermal and radiative conductivities for corresponding phases (Mouchovski, 2006). This model has been checked being a good approximation in case of $CaF_2$ single crystal growth at usually imposed BS thermal conditions (Mouchovski, 2007a) where the effective conductivities for both phases were estimated being approximately 20 times higher than their constituting thermal conductivities.

Analyzing further the inequality (14) it is seen the second term on its left-hand side, $(L_{molt}/2K_L)$, should have stabilizing effect upon interface stability; this effect is as much pronounced as the latent heat of fusion is larger and/or the thermal conductivity of the melt is lower. For crystals with semi-transparent melts where $K_{Leff} \approx K_{Lrad} >> K_{Lconc}$ the real contribution of $(L_{molt}/2K_{Leff})$ should be relevantly small.

### 3.4 Normal growth criterion

Any analysis of supercooling effect, causing transition from mono-crystalline to cellular growth for solid solutions with fluorite structure, will be correct only if normal growth proceeds on a rough surface. It is considered this mechanism being involved when the value of normalized latent heat of fusion $L_{molt}/kT_{m.p.}$ ($k$ – is the Boltzmann constant and $T_{m.p.}$ – the m.p. absolute temperature) stays below 2 (Jackson, 1958) or below 3.5 (Alfintzev et al., 1980). As shown in case this criterion failed, the growth proceeds by formation of two-dimensional nucleation. At that the high anisotropy of the process prerequisites both, the segregation coefficient and the surface energy for considered impurity, to depend strongly on growth direction. In this case besides cells' formation, a laminar distribution of impurity can occur caused by capturing of melt's layer adjacent to the growth surface.

## 4. Function of stability and properties for mixed fluoride systems

### 4.1 Mixed alkali earth fluoride systems

The alkali earth metal single fluorides $MF_2$ may form tri- or four-component solid solutions of type $Ca_{1-x}Sr_xF_2$, $Ca_{1-y}Ba_yF_2$ or $Ca_{1-x-y}Ba_xSr_yF_2$, which retain the cubic symmetry of the fluorite lattice. However, complete solubility of the starting single fluorides has proved to be possible only for the first system, the compositional phase diagram of which manifests a clear minimum for $0.4 < x < 0.5$ (Chern'evskaya & Anan'eva, 1966; Klimm et al., 2008; Weller, 1965) while the system's properties appear intermediate between those of $CaF_2$ and $SrF_2$ end members. Correspondingly the compositional dependence for stability function starts from and ends to zero K, being insignificant at the extreme minimum (Klimm et al., 2008). This crystal system has been recently an object of thorough investigation within a large compositional interval for specific case of simultaneous growth of batch of parallel boules with different $x$ where for $CaF_2$ was utilized fluorspar, concentrated to above 99.6 wt.% (Mouchovski, 2007b; Mouchovski et al., 2009a, b; Mouchovski, 2011).

Completely different behaviour at room temperature possesses $CaF_2$–$BaF_2$ system owing to a large difference in ionic radius between $Ca^{2+}$ and $Ba^{2+}$ and, accordingly, in lattice parameter between $CaF_2$ and $BaF_2$. As a result, after a primary crystallization of a continuous series of $Ba_{1-x}Ca_xF_2$ fluorite solid solutions with a minimum in the liquidus curve at $x < 0.5$, during the cooling there occurs a high-$T$ (at 900 °C) solid-state decomposition into a two-phase mixture of dilute $CaF_2$ and $BaF_2$-based solid solutions (Chernevskaya, & Anan'eva, 1966; Wrubel et al., 2006; Zhigarnovskii & Ippolitov, 1969). The phase diagram of this system has been recently reviewed by Fedorov et al. (2005) who showed it does not contain a continuous solid solution. Thus, on cooling the $70CaF_2$–$30BaF_2$ melt (mol%), a $Ca_{1-y}Ba_yF_2$ solid solution ($a$-phase) solidifies as a primary phase, which at (1039 ± 5) °C reacts with the melt to form a second fluorite α-phase, $Ba_{1-x}Ca_xF_2$. This $a$-solid solution decomposes on further lowering the temperature into two phases: $a_1$ and $a_2$. When the temperature reaches (870 ± 5) °C, it comes up to a eutectoid equilibrium between three fluorite phases: $a_1$, $a_2$, and $\beta$. This equilibrium is broken at lower temperature since quenching at a rate of 250 °C/s cannot retain the $a_2$-phase and it is decomposed. As a consequence of these thermal events, the resultant composite material consists at room temperature of two fluorite phases, $a_1$ and $\beta$. Surprisingly, it posses a unique ionic conductivity that is respectively 25 and 330 times higher than those of the parent $BaF_2$ and $CaF_2$ – phenomenon that was explained in terms of the electrical properties of the material's interfaces (Maier, 1995). Materials with such properties appear of great interest for solid-state ionic opening up new opportunities for engineering medium temperature fluoride ion conductors (Sorokin et al., 2008).

Nevertheless, the mixed AEM fluoride crystals have been studied mostly for their promising lasing properties after doping/co-doping by appropriate trivalent Ln-ions with eventual charge-compensation by univalent alkali ions. For $Ln^{3+}$-based lasers these properties depend strongly on both the symmetry and strength of the crystal field so they may be purposely controlled by appropriate modification of the host material.

### 4.2 Mixed alkali earth – Rare earth fluoride systems

As denoted by Fedovov at al. (1988) the phase diagram and resulting function of stability for any $MF_2$–$REF_3$ system is in accordance with the type and sizes of RE ionic radius that, in turn, reflects on relevant interdiffusion coefficient. The authors established a regular dependence of $F(x)_{crit}$ on RE ionic radius, with a clear minimum for M = Ba and Sr, corresponding to transition from $k_o > 1$ for the larger cations in the beginning of the Ln-series to $k_o < 1$ for the smaller cations pertaining to the yttrium sub-series. In case of M = Ca any regularity was not found due, probably, to fine partitioning of the Ln-series on several sub-series (Djurinskij & Bandurkin, 1979). Further calculations led the authors to interdiffusion coefficient for La (the RE with largest ionic radius), $D_{La}$ into fluorite matrix being independent on partition of $LaF_3$ second component when its concentration exceeded 0.16 mol.%. At the same time $D_{RE}$ was shown to keep a linear dependence on RE-radius, decreasing approximately twice from La to Yb in the range of $10^{-5}$ $cm^2/s$.

Extending broadly the investigation of $MF_2$–$REF_3$ systems, Kuznetzov and Fedorov (2008) established empirically a clear maximum in solidus compositional dependence of $k_o$ for $SrF_2$–$CeF_3$ system, expressing it by power function equation, the coefficients of which appear complex functions on data extreme. The relevant $F(x)_{crit}$, showing a slight maximum (10 K) at $x_S = 0.1$ mol.% and nearly zero minimum at $x_S = 0.25$, goes up hyperbolically attaining so high values at $x_S > 0.5$ that excluded, in practical the formation of solid solution crystals. The same specificity for $F(x)_{crit}$ dependence on $x_S$ were found out for Nd, Pr and La whereas for the rest REs the authors did not reported of any maximum at low RE content. They announced similar results in case M = Ca where $F(x)_{crit}$ started to increase hyperbolically without any intermediate maximum, being less pronounced for Ce whereas for La $F(x)_{crit}$ tended to linearity as for $x_S < 0.5$ its values stayed firmly below 25 K. In case M = Ba, again for Ce, Nd, Pr and La it was established the specific for M = Ca maximum - minimum - hyperbolical increasing succession of $F(x)_{crit}$ dependence where the maximum variations were larger when being displaced towards the lower $x_S$. For the rest RE significant alterations in curves' course were obtained that suggests formation of variety of complex structures consisting in large Ba ions and much smaller, different in sizes RE ions. To the same suggestion leads the comparison of $F(x)_{crit}$ vs. RE radius dependence for Ca and Ba respectively; one can see a clear minimum for M = Ca, displaced to larger radiuses for higher concentrations whereas the minimum turns out considerably broadened when M =

Ba. Accordingly to these results one may anticipate specific behavior of the key physical-chemical properties for studied mixed fluoride systems grounded on their structural peculiarities. Here of particular importance appears the thermal conductivity *T*-dependence mostly used for assessing the eventual application of particular $MF_2$–$REF_3$ systems for creation of highly effective solid-state lasers, with $RE^{3+}$ as activator.

Nowadays it is apprehended $M_{1-x}Yb_xF_{2+x}$ systems with $x < 0.5$ being with priority as comparing corresponding $M_{1-x}Nd_xF_{2+x}$ systems since the usage of $Yb^{3+}$ instead $Nd^{3+}$ as ion-activator in laser materials leads to several approved advantages: *i*) simplicity of the structures of the laser levels for ytterbium; *ii*) smaller difference between the wavelength of pumping and generation of pulses; *iii*) lowered heat losses; *iv*) wider bands of the emission spectrum that is especially convenient for generation of short pulses and creation of tunable lasers with larger lifetime of the upper excited level. Here again the usage of $CaF_2$ for the matrix seams most promising and combination $CaF_2$–$YbF_3$ leads to formation of heterovalent fluoride solid solution $Ca_{1-x}Yb_xF_{2+x}$ at $x \leq 0.41$ (Ito et al., 2004; Lucca et al., 2004; Sobolev, & Fedorov, 1978). The thermal conductivity *T*-dependence for this system has been investigated thoroughly by Popov and co-workers (2008a) that established specific alteration in conjunction with considered compositional interval. At the lowest compositions, $x \leq 0.001$, a decrease of thermal conductivity with increasing *T* was observed within 50-300 K, that is, up to room temperature. Such behavior is typical for crystalline materials and corresponds to a decrease in the mean free path of phonons with temperature. For $x$ within 0.005–0.03, smooth maximums were noted in the curves that turned out displaced downward into the range of high *T* with increasing $x$. Importantly, in the low *T*-interval there occurs a transition from behavior typical of single crystals (with a maximum in *T*-dependence) to behavior characteristic for glasses, with a monotonic decrease in thermal conductivity with *T*-decrease. As firmly considered such transition to glasslike structures is connected with formation, accumulation, and agglomeration of clusters of oppositely charged defects, which leads to disturbance of short-range order with retention of long-range order [Fedorov, 1991, 2000; Kazanskii et al., 2005; Sobolev et al., 2003). Within the next compositional interval, $x = 0.09$–$0.25$, the thermal conductivity, was found out to grow up monotonically with *T*, which is typical behavior for any disordered systems but is completely inadmissible for crystals of simple stoichiometric compounds. Similar glasslike behavior of any concentrated heterovalent solid solutions $M_{1-x}RE_xF_{2+x}$ of $REF_3$ into fluorite-type compounds $MF_2$ (M = Ca, Sr, Ba, Cd, Pb) was noted repeatedly (Fedorov, 2000) for the thermal conductivity as well (Popov et al., 2008b, c). Here unifying result appeared the mean free path for phonons (within 0.09–0.25 compositional interval) was found only weakly *T*-dependent, becoming on the order of unit-cell dimensions (Popov et al., 2006). This fact suggests "growth-in" of clusters into the fluorite-type lattice and appearance of nanoheterogeneity in the relevant solid solutions.

## 5. Purpose and aims: Control upon interface stability and crystal quality

It can be distinguished two types of factors, influencing the stability of growing interface: substantial factors that refer to peculiarities of growing compounds, determining liquidus–solidus phase diagram, and technological (apparatus) factors related to configuration of the thermal field, established into the load, and the real CR. Nevertheless the technological factors are in subordinated position to the substantial ones since the proportion of starting compounds predetermines often the region of permissible alterations for the imposed thermal conditions and the speed of crucible withdrawal.

The present survey deals with different approaches for providing an efficient control upon morphological and geometric stability of the moving CF, keeping its shape near to planar during the growth of particular single or mixed fluoride systems. The accent is being put on single $CaF_2$ and $Ca_{1-x}Sr_xF_2$, where both cases natural fluorite (fluorspar) is being used as starting material. The influence of several different impurities (eutectic particles of non-soluble compositions) or ionized molecules (oxygen-containing anions or cations embedded into the lattice) upon interface stability is analyzed semi-quantitatively for single $CaF_2$ by using simple supercooling criterion. Here the aim is how to be implemented an efficient control on trace levels of residual impurities into starting fluorspar and its melt as well as on other contaminants that enter eventually the CZ during crystallization itself.

The same supercooling criterion is being utilized for investigating the stability function in case of simultaneously growth of $Ca_{1-x}Sr_xF_2$ boules with different composition. Here the exact knowledge of liquidus–solidus phase diagram is rather important in particular, when for $CaF_2$ end member is being chosen two types fluorspar, differing each other with RE impurities' content (total amount below 100 ppm). Thus specified phase diagram (with expected decrease of the initial negative slope and displacement of the minimum towards the lower Sr-content $x$ (Mouchovski, 2007a) should alter relevantly the effective segregation coefficient, which equilibrium value appears a structurally sensitive characteristic of the mixing processes. Here aim is determination of the compositional dependence for $k_o$ that will offer scope for hypothesizing about possible mechanisms for incorporation of the second component - Sr ions with $CaF_2$ matrix (first component).

The phase diagram character influences decisively as well the interdiffusion coefficient, so that the determination of its compositional dependence is another aim of the study.

The optical characteristics of thus grown crystals - as most sensitive properties of the medium - are measured, aimed they being related to corresponding values for stability function for comparative analysis to be implemented.

As general goal of the survey appears determination of appropriate interval for crucible speed towards the cold furnace zone, within which a normal growth to proceed at minimal supercooling effect in conjunction with the established real temperature distribution into the load. This means to be implemented control upon the technological factors so that: *i*) to be attained controllably sufficiently steep axial (vertical) $T$-gradient into the furnace unit (FU), keeping at that a minimal radial $T$-gradient into the load whereat a planar or slightly convex $IF_{L/S}$ to be maintained; *ii*) to be set up an appropriate SC, that to be very close to the real CR along the mostly crystal length.

## 6. Methodology

Complex growing methodology is applied to provide minimum contamination into CZ, and effective control on the thermal condition into the FU.

It is implemented efficient deep preliminary purification (PP) of the starting fluorspar for substantial reduction of mostly metal cations (except those pertaining to Ln-group) and some metal oxides resulted from decomposition of accessory minerals accompanying the basic fluorite. The PP includes consecutive chemical treatment of grained fluorspar portions into HCl and HF acids followed by high-$T$ treatment in the presence of melt scavenger (up to 2 wt.% $PbF_2$ or/and $ZnF_2$ additives) so that the purity of the produced polycrystalline/sintered precursors attains level of 99.6 wt.% (Mouchovski et al., 1999; Mouchovski, 2007a).

Crystal growth by BS technique, where the constant speed of crucible withdrawal is step-wise changed between 0.2 and 0.6 cm per hour, is accomplished into two types' multicameral crucibles, the specific construction of which assist for restricting considerably penetration into CZ of undesired oxygen/water vapour ionised molecule from the vacuum chamber environment. Specifically, the sizes of the openings (channels) into inner cameras' lids are adjusted of order of the mean free path of gaseous ions so that their back movement inside the cameras being under control by Knudsen diffusion (Mouchovski, 2007a).

Both types' multicameral crucibles consist of central nest surrounded by several axi-symmetric peripheral cameras (Mouchovski et al., 1996). The first type crucible ("Tube support") contains 9 cameras-inserts (with diameter of 2.48 cm), placed in parallel and close to each other in peripheral tubular compartment. The free spaces around and above the inserts are, essentially, additional mass-transport resistance as regards all gaseous species penetrating from vacuum environment. Opposite to that, the 8 cameras (with diameter of 2.56 cm) of the other type crucible ("Revolver support") are surrounded my solid graphite mass, the relatively large thermal conductivity of which ensures the radial heat flux throughout the load (graphite walls surrounding crystallizing molten portions of fluorspar and several connected free spaces) being correspondingly released.

The thermal field in such complex load follows the set $T$-programs for furnace power supply of tailor-made Bridgman-Stockbarger Growth System (BSGS). Its FU consists of close package graphite screens wherein upper and lower heaters differentiate hot (Z1) and cold (Z2) zones, respectively, separated by diaphragm (SD), differentiating relatively long adiabatic zone (AdZ) wherein the radial heat-exchange should expect being insignificant. Supplementary, system of molybdenum shields, part of which is related to moving crucible, has been introduced into the FU. The altering configuration of this system allows much more precise regulation of the thermal field into the FU and into the load, respectively (Mouchovski, 2007a). The details are discussed in sub-section 7.2.

Products of crystal growth are batch of simultaneously grown boules wherefrom are prepared pairs of parallel optical windows, taken from one and the same sections of the cylindrical body of the boules. The windows are finished according to requirements for laser grade $CaF_2$ (Mouchovski et al., 2011) that reduces considerably the reflectivity radiation losses from the parallel planes surfaces within and below UV region. Approximately equal distance (0.6–1 cm) between windows' pairs for each particular batch of boules allows implementation of reliable comparative analysis of composition and characteristics of the grown crystals.

The windows are utilized for determination of: *i*) Ca-proportion, (1-$x$), in the final solid solutions by using of X-ray diffractionless analyzer BARS-3; *ii*) light transmission $t$-spectra, measured within UV-IR range by means of high-sensitive spectrophotometer, a Varian Cary type 100. Series of $t$-measurements are also obtained at discrete wavelengths ($\lambda$ = 248.3 nm, 510.6 nm and 6.45 μm) by applying so called Valour Lasers Irradiation technique (VLIr)

(Mouchovski et al., 2011). The corresponding light extinction losses per unit optical path are calculated from formulas grounded on Lambert-Beer law approximation where $t$ is presented as a sum of absorption and scattering parts and window's plane surface reflectivity is being excluded.

The stress-induced birefringence in the windows is determined by means of polariscope/polarimeter PKS-250.

The concentration and distribution of residual impurities along the height of grown boules is determined applying Atomic Absorption Spectroscopy (AAS), Solid Sampling Electrothermal Atomic Absorption Spectrometry (SS-ETAAS) and Neutron Activation Analysis (NAA) - for the RE elements - techniques (Detcheva & Hassler, 2001; Detcheva & Havezov, 1994, 2001, 2005, as cited in Mouchovski, 2007).

For assessing the CF position, $x_{CF}$, along the FU it is applied originally developed indirect technique, which involved determination of a Quenched Interface (QI) in multicameral crucible loaded by portions of concentrated grained fluorspar (Mouchovski et al., 1996a).

Empirically derived formulas, given elsewhere (Mouchovski, 2007a), relate $x_{CF}$ to $x_1$-variable (expressing crucible movement from the starting position) and the changes in the set $T$-regimes of furnace heaters. The origin of the basic axial coordinate, $z$, lies on FU cross-section separating AdZ from Z2. The height of crystallizing part of the boules is determined by the difference ($x_{CF}$ - $x_{con}$) where $x_{con}$ defines the distance passed by the plane section of the conical cameras' tips. All variables are reduced appropriately to take dimensionless form. The boules' height $h_{boule}$ is expressed in $x_1$* units as partition from crucible withdrawal for a given run by formulas:

$$x_1^* = (z_{incon} - z)/l_{crmov} \tag{15}$$

and correspondingly:

$$z = z_{incon} - l_{crmov} \times x_1^* \tag{16}$$

where $z_{incon}$ is the initial position of cameras' tips wherefrom the crystallization begins to propagate at the chosen starting position of the crucible into Z1 and $l_{crmov}$ represents the distance whereto the crucible moves downward into the FU.

The real $T$-gradient along the FU is assessed approximating the calculated axial $T$-gradient obtained after differentiating the axial $T$-profile of the furnace, measured in empty crucible (Mouchovski et al., 1996).

The real CR is estimated as the set SC is being corrected accordingly the varying shift in $IF_{L/S}$ position during crucible movement.

The liquidus - solidus curves for particularly studied $Ca_{1-x}Sr_xF_2$ system (Bulgarian fluorspar as starting material) are obtained by DTA measurements accomplished using a Stenton Red Croft STA instrument in argon flow and using graphite crucibles. The measured points lie within the experimental error on curves estimated by using formula, adopting recently specified phase diagram for the system in case both end members are chemical compounds [see Fig. 3 in Mouchovski et al., 2011]:

$$T_{liq/sol}(Mouh) = T_{liq/sol}(Klimm) - (1-x)\,\Delta T_{x=0} \qquad (17)$$

Here a linear decrease is being approximated for calcium proportion (1–$x$) with the initial $CaF_2$ m.p.-difference, $\Delta T_{x=0}$, equal to 1691 – 1648 = 43 K for the used types of fluorspar.

Thus obtained datasets are fitted by 8-order polynomial with correlation coefficient above 0.99.

The critical values for stability function $F(x)_{crit}$ are calculated from the phase diagram where both dependences, $m(x_o)$ and $\Delta x_o = x_S - s_L$ are indirect functions of $T$.

The compositional dependences of segregation coefficient and interdiffusion coefficient are calculated from (8) and (13) being in conjunction with found alterations of the real $T$-gradient and real CR during the growth of each particular boule.

## 7. Experimental

### 7.1 Interface stability during the growth of $CaF_2$ crystals

*Effect of oxygen-containing contaminants on interface stability*

In case of $CaF_2$ crystal growth by using fluorspar, the melt of which is contaminated by oxygen-containing molecules/ions, a sequence of physical processes and chemical reactions leads to eutectics formation since the produced final compound - CaO - is not isomorphic with $CaF_2$ and cannot dissolve into the lattice to form a solid crystal solution.

The eutectics formation may consider on the grounds on supercooling effect arisen. For the purpose the right-hand side of criterion (10) is presented by the crucial $T$-gradient along the layer ahead the CF with thickness $\delta_i$, whereat the equilibrium between the average interdiffusion coefficient for particular impurity "$i$", $D_i$, and the maximum linear CR, $V_{cryst}$, starts to disrupt:

$$\Delta T_{cru}/\delta_i = (RT_{cr}^2/L_{molt})[x_{o(b)}(k_{eff}^{-1} - 1)](V_{cryst}/D_i) \qquad (18)$$

Substituting the proportion between the average diffusion coefficient, $D_i$, and $V_{cryst}$ for $\delta_i$ in (18), the decrease in crystallization temperature $T_{crys}$, due to enrichment of the layer ahead the CF by particular impurity, is expressed by $\Delta T_{crys}$-quantity according to formula:

$$\Delta T_{cru} = (RT_{cryst}^2/L_{molt})[x_{o(b)}(k_{eff}^{-1} - 1)] \qquad (19)$$

In case $CaF_2$ crystal growth proceeds where the used fluorspar possesses m.p. (1648±5) K and $L_{molt}$ = 7100 cal/mole, then $(RT_{cryst}^2/L_{molt})$ = (760 ± 2) K. Thus, when the principal contaminant is insoluble CaO with $k_{eff} \ll 1$, it is fulfilled $(k_{eff}^{-1} - 1) \gg 1$. At this junction even at relatively small CaO concentration, the product $[x_{o(b)}(k_{eff}^{-1} - 1)]$ may attain significant value, exceeding of two order of magnitude $x_{o(b)}$. Using arbitrary: $k_{eff} \approx 0.01$ and $x_{o(b)} = 5\times10^{-4}$ mole parts, the calculated $T$-fall is significant: $\Delta T_{crys} \approx 39$ K.

On the other hand, $D_i$ for $O^{2-}$ dissolved in $CaF_2$ melt is too low ($\approx 10^{-11}$ m$^2$/s) so that these anions are immobile in practical compared to $F^-$ anions, which, possessing nearly the same effective ionic radius, bear twice less negative electrical charge. At this junction, within the usually attained interval for $V_{cryst}$ that follows approximately the SC (2–10 mm/h), $\delta_i$ stays of order of $10^{-3}$ cm (10 μm) while the crucial $T$-gradient, $\Delta T_{cru}/\rho_{im}$, is calculated equal to $3.9\times10^4$ K/cm (39 K/μm). Such extremely steep positive $T$-gradient along the enriched to CaO layer ahead the CF should exclude mono-crystal growth to proceed. Instead, unsteady thermodynamic conditions will initiate single crystal growth from the heterogeneous region of $CaF_2$ - CaO phase diagram. As a result, it appears a great number of fine fluorite crystals with higher temperature of crystallization than $T_{crys}$ at the CF, which produce an intensive light-scattering and the grown crystals turn out white-milk in colour, being fully opaque within UV-VIS range.

To stabilize the growing equilibrium in accordance with criterion (18), one can lower appropriately $V_{cryst}$, that is, the SC. However its excessive decrease turns out unacceptable for industrial crystal production and, besides, may cause considerable evaporation and/or decomposition of the molten material with following break in stoichiometry owing to loss of $F^-$ from anionic sub-lattice. Thus for eliminating the CMSc due to the presence of CaO, mostly oxygen contaminants, involved in Ca oxidation, have to be removed before $T$ to exceed 880 °C whereat temperatures the rate of chemical oxidation to increase dramatically. Alternatively, the solubility of mostly oxygen-containing contaminants in the solid phase may be risen up considerably by introducing definite amount of tri-valence RE ions into the lattice since their extra-charge leads to local compensation of the negative charge of $O^{2-}$ and they turn out incorporated into anion sub-lattice. Combining the two methods discussed, one may anticipate the established constitutional $T$-gradient to remain satisfactory low so that it may be efficiently compensated by ensuring opposite steep axial $T$-gradient along the FU and by setting moderately low SC.

According to criterion (18), the presence of RE or any other metal impurities, the effective segregation coefficient of which remains closed to unity, should not contribute for enhancing the CMSc effect in case their concentration does not exceed several hundreds of ppm in starting fluorspar whereat they could not be treated as second solid solution component. However, in many cases when UV-grade $CaF_2$ has been grown, the RE impurities turns out rather undesired being embedded into the lattice since they act as optical active centres of specific light absorption. At this junction, not embedment into the lattice, but efficient removal of these impurities from the CZ appears obligatory for growing crystals with needed optical characteristics.

It should be taken into consideration that melt supercooling, not related to the purity in CZ, can arise being caused by appearance and constant rise up of radiation flow throughout the growing transparent crystal, surpassing increasingly the conductive flow towards the cold furnace zone.

## 7.2 Interface stability during the growth of $Ca_{1-x}Sr_xF_2$ crystals

### 7.2.1 Growing runs

Two growing runs are performed by using correspondingly the two types of multicamera crucibles placed in BSGS thermal field determined by two limiting cases as regards the mutual disposition of the movable part (batch of rings fixed on crucible stem) and fixed part (a thin liner, 15 cm long, mounted through the bore of SD, 2.47 cm thick) of the MoShS. Thus, 9 rings participate in *run1* whereas no rings are slept on the stem in *run2*. Besides the SC is chosen different for the two runs altering in step-wise manner (between 2 and 3 mm per hour - for run*1*) or being constant (equal to 6 mm per hour - for run*2*) – **Fig. 1**. Such SC regimes are thought being in conjunction with the two limiting cases for MoShS effect. These cases concern total re-distribution for heat flux throughout the load. It is supposed the liner to play a role of a long pseudo-diaphragm that restricts the radial heat losses, reflecting the emission from both, the lateral surface of moving crucible and the lower furnace heater. On the other hand the fixed rings below the bottom of moving crucible cause - at a given instant - a jump in the current gradual alteration of the vertical (axial) to radial heat exchange, depending on crucible position into the FU. Initially, when crucible with the number of rings on its stem is being fixed on the starting position into Z1, all rings turn out within the liner region (LR). At this junction, after stabilization of power supply, some amount of heat will be accumulated in between crucible bottom and the upper batch's ring that causes an increase in temperature into the load, thus providing more frugal melting of the charged portions. Further, by withdrawing the crucible towards Z2, the rings will pass consecutively outside the LR, whereat the lateral heat transport from the load increases. That way, the total heat flux throughout the load turns out facilitated, leading to *T*-drop down therein whereat the CF will shift downwards correspondingly. At the absence of rings below crucible bottom such re-distribution of the total heat flux should not occur and only the liner will affect the CF position.

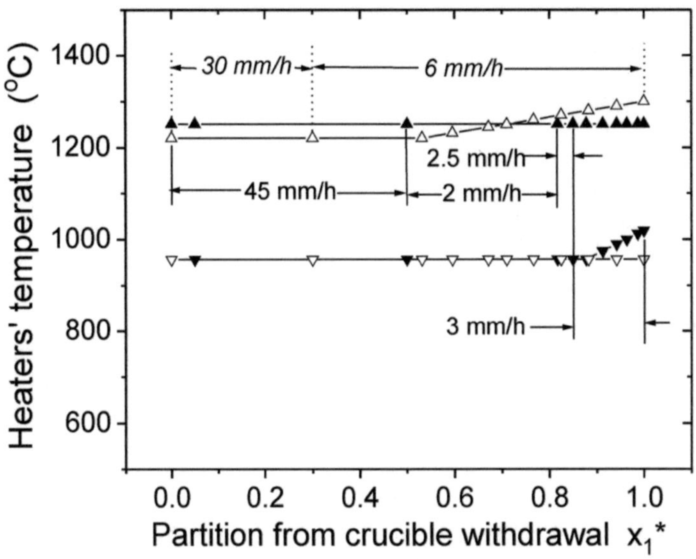


Fig. 1. Temperature regimes for BSGS-furnace zones during the growth of calcium-strontium fluoride mixed crystals: run 1 (2-2.5-3 mm/h): (▲) - $T_1$(Z1), (▼) - $T_2$(Z2); run 2 (6 mm/h): (△) - $T_1$(Z1), (▽) - $T_2$(Z2). The intervals for set crucible speed are marked as well.

The initial proportions of $CaF_2$ to $SrF_2$ are controlled so that $x$ in the grown $Ca_{1-x}Sr_xF_2$ solid solutions to vary within 0.007 and 0.307 (run*1*) and 0.383 and 0.675 (run*2*).

### 7.2.2 Phase diagram

The $Ca_{1-x}Sr_xF_2$ phase diagram when the used fluorspar (concentrated to 99.6 wt.%) contained some insoluble amounts of silicon, aluminium and iron oxides and traces of other metal impurities where the RE varied up to 100 ppm, is shown on Fig. 2. It is seen both liquidus and solidus curves pass through aezotropic minimum at $x \approx 0.35$ sifted to lower Sr concentrations and considerably low temperature ($T_{Lid} \approx 1620.9$ K) as comparing az.p. $x \approx$

0.42 ($T_{Lid} \approx 1645$ K) - established by Klimm and co-workers (2007) utilizing pure chemical compounds for the end members (see Fig. 3 in Mouchovski et al. (2011)).

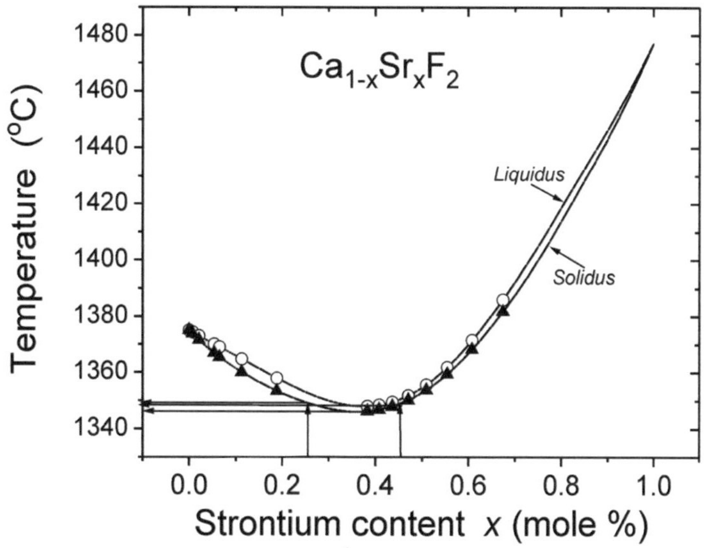


Fig. 2. Binary system of $CaF_2$ - $SrF_2$ where for $CaF_2$ end member is used fluorspar, mined from Slavyanka Bulgarian fluorite deposit, with m.p. (1375 ± 10) °C; $SrF_2$ is "suprapur" quality (supplied by Merck), with m.p. 1477 °C. For carried out growing runs by arrows are shown: $\Delta T_{sol}$ variation of 2.8 K around the aezotropic point of 1619.9 K for $\Delta x$ within 0.25-0.45, ($x = 0.352$ at az.p. minimum). The corresponding $\Delta T_{liq}$ variation of 5.7 K lies within 1625.6 and 1619.9 K.

At this junction relevant alterations in calculated compositional dependences for critical stability function $F(x)_{crit}$, interdiffusion coefficient $D_{Ca/Sr}$ and equilibrium segregation coefficient $k_o$ will occur. They are grounded on theoretical expression (13) that relates $D_{Ca/Sr}$, $F(x)_{crit}$, and $V_{cryst}/G$-ratio wherein - as discussed below - the real $T$-gradient ahead the CF may change considerably with crucible withdrawal into a thermal field with altering axial $T$-gradient. At the same time the real CR differs from the set SC depending on the shift in CF-position $x$. Moreover, as shown experimentally elsewhere (Mouchovski et al., 2011) the crystallization in each camera with different composition for loaded mixture (different $x$) will start at different crucible position $x_1$*.

As regards $k_o$, according to formula (7), it depends in complex way indirectly on $x$ via $L_{melt}(x)$ and $T_{melt}(x)$ functions.

### 7.2.3 Equilibrium coefficient of segregation

This is important parameter since indicates how the differences between molar concentrations of liquid and solid phases for the second (Sr) component at the $IF_{L/S}$ alter in accordance with mixtures' composition, $x$. As seen (Fig. 3) the calculated curve cross the basic line $k_o = 1$ at $x \approx 0.35$ indicating that compositions not far from this value appear especially appropriate for growing to proceed at minimum CMSc effect. Nearly constant slope of the curve is observed within $0.13 \leq x \leq 0.58$ whereas at higher Sr content $k_o$ stays in practical uniform, varying slightly around 1.2, thus revealing some rejection of the second components ions by moving CF. At infinitively low Sr-concentrations $k_o$ declines steeply to limiting value of ≈ 0.79.

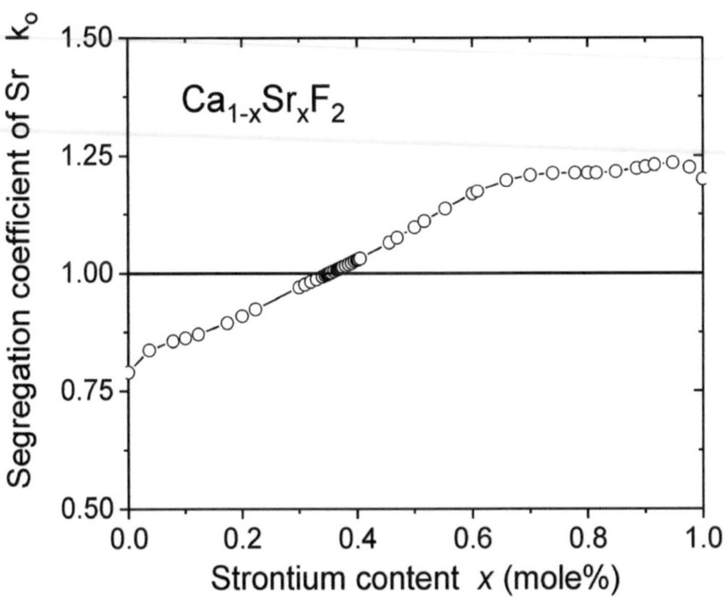


Fig. 3. Equilibrium segregation coefficient of Sr as a function of its content in simultaneously grown batch of $Ca_{1-x}Sr_xF_2$ boules where fluorspar (m.p. of (1648 ± 5) K) is being used for $CaF_2$ end member.

### 7.2.4 Compositional dependence of stability function

As suggested by the view of the phase diagram on Fig. 2, the compositional dependence of stability function (Fig. 4), calculated on the base of run*1*+run2 dataset (17 points), reveals a clear minimum close to zero at $x \approx 0.47$ that means, the corresponding run*2*-boule should be grown at mostly favourable conditions as regards CMSc effect. For two other boules, where $x$ is positioned not far from the minimum (0.383 and 0.55, respectively), $F(x)$ remains below 1.6 K that can be considered being acceptable reduction of the CMSc effect. Only boules with extremely low $x$ or $1\text{-}x$ should be at similar favourable growing conditions.

Surprisingly, the established $F(x)$ minimum at $x \approx 0.47$ differs from az.p. at $x \approx 0.35$. The reason for such divergence lies in the fact that the critical values for stability function depends as well on the partition from crucible withdrawal $x_1{}^*$ via the dependence on this variable for the ratio of axial $T$-gradient to CR, multiplied by $D_{Ca/Sr}$. In turn $D_{Ca/Sr}$ depends on $x_1{}^*$ via the opposite ratio for these apparatus/runs factors, multiplied by $F_{crit}(x)$.

The complex functionality for both dependences requires their investigation as regards the two runs growing conditions.

### 7.2.5 Determination of the real axial T-gradient into the load

The magnitude of the real $T$-gradient ahead the CF, $G_{CF}$, depends on CF-shift upward or downward during crucible movement through the FU $T$-field, characterized by its axial $T$-

profile and relevant furnace gradient $G_{fur}$. Besides, the effective thermal conductivity for both solid and liquid phases and the interface transition resistance of the interface affect $G_{CF}$. At this junction a semi-empirical formulas, giving the alteration of these *T*-gradients are much better to be derived for each particular FU configuration instead to utilize speculatively any sophisticated thermal model developed for the purpose.

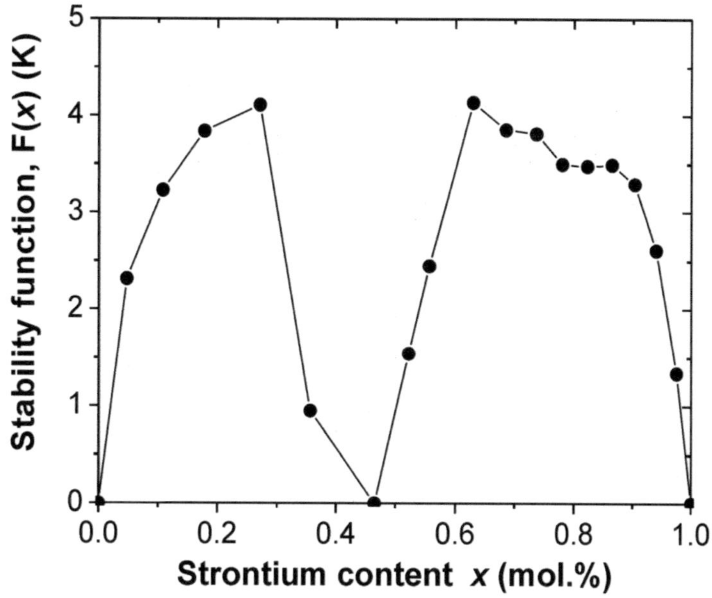


Fig. 4. Stability function *F(x)* for $Ca_{1-x}Sr_xF_2$ system with $CaF_2$ m.p. 1648 K.

The first step here is determination of the axial *T*-furnace profile measured in empty crucible by moving downwards a high sensitive thermocouple and keeping a constant distance between its junction and the inner surface of crucible bottom. The power supply is being *T*-controlled by means of highly precise programmers/ controllers. The thermal field along the FU is established according to set MoShS configurations. The differentiated *T*-profiles (Fig. 5) reveal specific alteration for $G_{fur}(x_1^*)$ slopes along the FU. As seen for run*1*-conditions the slope of curve *1* rises up gradually. This is a result of the obtained parabolic shape for initially measured *T*-profile, which lowering branch is formed at the consecutive passage of the rings outside the LR. The role solely of the liner is apprehended as comparing curve *2* (no rings) to curve *1* (9 rings); the less steepness of curve *2* manifests the absence of rings results in facilitating the heat flux towards Z2. Nevertheless the slope seams being with sufficient steepness, attaining a constant value of 13.7 K/cm just below the AdZ, that to ensure stable, favourable conditions for normal growth to proceed.

Besides the slope, the absence of rings affects as well the position for *T*-profile that appears shifted more than 100 K to the lower temperatures as comparing the *T*-profile for curve *1*. That means the crystallization should start at much higher position of the load - into Z1. To compensate such loss of heat into the load a relevant increase of power supply is being carried into effect (Fig. 6a, b) so that the growth to start and proceed more favourably in LR.

The correct assessment of CMSc effect requires $G_{real}$ on $G_{fur}$ dependence to be followed along the height of growing boules. First of all that means to be clarified the regularities those govern the CF position during the growth itself. As seen (Fig. 6a, b) the CF-position depends substantially on the composition of solid solution crystals. Thus for covered interval, $0.007 \leq x(Sr) \leq 0.307$, run*1* conditions determine the CF-positions being either entirely in Z1 ($x$ = 0.007), that means convex CF shape, or to vary around the lower (boundary) cross section of the AdZ ($x$ = 0.307) whereat the relevant shape will tend to become slightly concave.

Run2 conditions ensured much larger alteration in CF-positions - from entirely above Z2 ($x$ = 0.675) to mostly in Z2 ($x$ = 0.383). The best appropriate composition seems to be for the crystal with $x$ = 0.55 (Fig. 6b) where the CF changes gradually and insignificantly, being entirely within the AdZ that presupposes its shape to be kept nearby planer.

On each one CF-position into the FU corresponds particular slope of both, the set furnace *T*-gradient $G_{fur}$ and relevantly established real *T*-gradient ahead the CF, $G_{real}$. Thus it can be followed the alteration in $G_{fur}$, respectively $G_{real}$, during the growth of a boule with definite height, $h_{boule}$, presented in dimensionless form: $h_{boule}^* = h_{boule}/l_{crmov}$.

For the accomplished two runs: $z_{incon}$ = 11 cm while $l_{crmov}$ = 17.7 cm (run*1*) and 20.6 mm (run*2*).

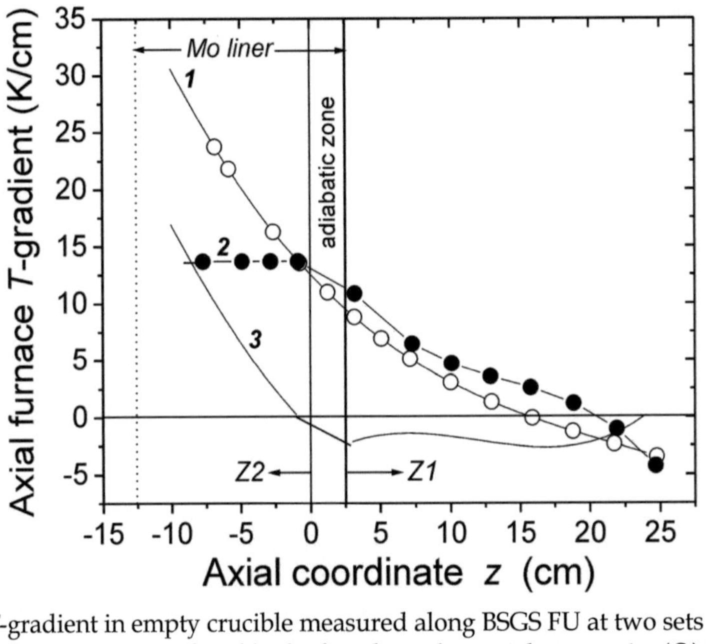


Fig. 5. Axial *T*-gradient in empty crucible measured along BSGS FU at two sets of thermal conditions. The data points are fitted by high order polynomials: curve 1 - (○) liner + 9 rings on crucible stem; curve 2 (●) - liner without rings Curve 3 – the difference between curve 1 and curve 2.

At boundary values for Sr-content $x$ in the grown solid solution system - boules 0.007 and 0.307 (run*1*) and 0,383 and 0.675 (run2) - the maximum, mean, and minimum $G_{fur}$ ($G_{real}$) are given in Table 1 while the *T*-gradient alterations are shown in Fig. 7. The initial analysis is grounded on the average values for $G_{real}$ (column 7 in Table 1) as the maximal deviation is being estimated using the one and the same reduction factor of 2.8 for reduction of $G_{fur}$

within the LR that cover AdZ and the upper section of Z2. Thus reduced gradient is considered to stay constant within a particular section of the LR. The average $G_{fur}$-values are given together with their maximum deviation.

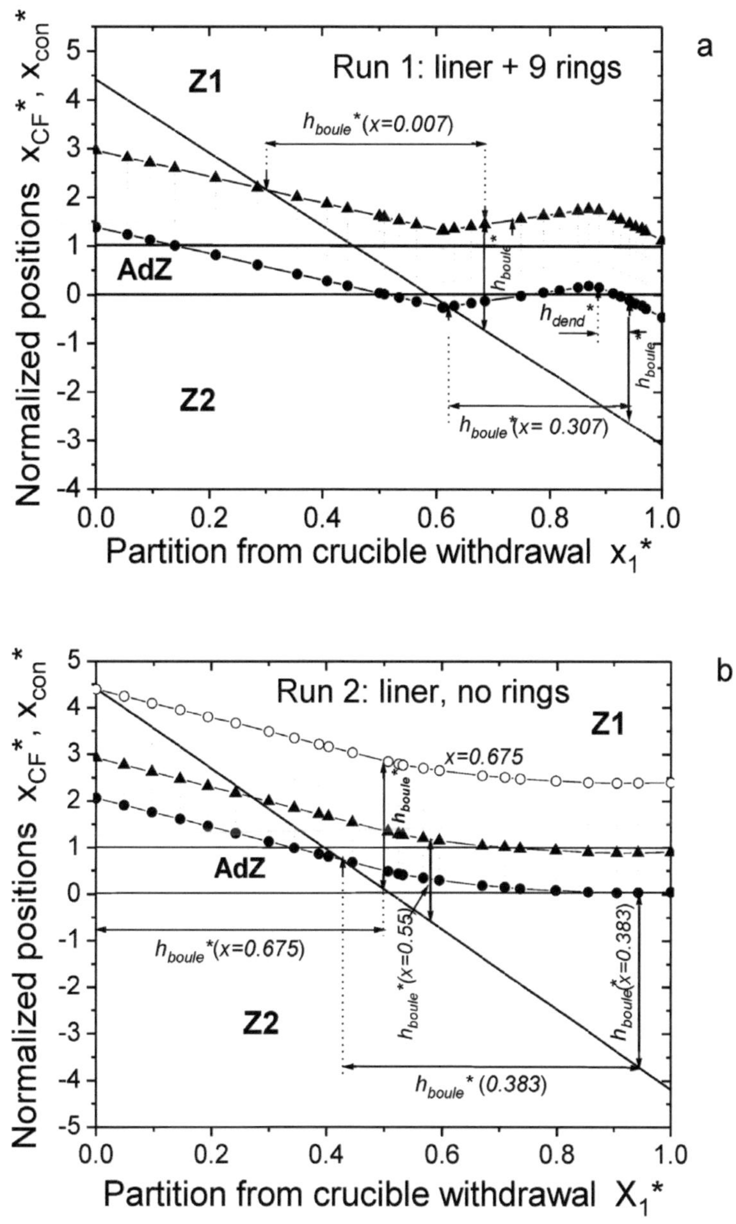


Fig. 6. a, b. Normalized positions of the CF and cameras tips' cross-section $x_{con}$* along the FU depending on the partition from crucible withdrawal during two growing runs of $Ca_{1-x}Sr_xF_2$ boules where Sr content $x$ varied between: **a** (run 1) – 0.007 (▲) and 0.307 (●). **b** (run 2) – 0.383 (●), 0.55 (▲) and 0.675 (●). The lines describing $x_{con}$* are given appropriately. The total boules' height is shown, normalized by AdZ thickness, $h_{boule}^* = h_{boule}/l_{Adz}$.

Comparing the alterations in $G_{real}$ for the two runs (Table 1 and Fig. 7) one can see divergence from $G_{real}$(Av) for run2-boules that attains 53% but only for those boules wherein the Sr appears clearly the dominant component, since then the boules should be grown at permanently increasing *T*-gradient. At lowering the *x* around and below 0.5 the CF shifts in a manner the growth to proceed more and more at the flat *T*-gradient section so that the divergence from the average falls down rapidly attaining the insignificant 12% at *x* minimum of 0.383. For this run $G_{real}$ should alter between 4.8 and 11.3 K/cm ($x = 0.55$). On the other hand the divergence for run*1*-boules varies between 27% ($x = 0.307$) and 42% ($x = 0.007$) but the first value is related to the lowest $G_{real}$ average for this run of only 6 K/cm.

| No. run | x(Sr) | $x_1$* (from/to) | $G_{fur}$(min/max) (K/cm) | $G_{fur}$(av) (K/cm) | CF-position CF-shape | $G_{real}$ (K/cm) |
|---|---|---|---|---|---|---|
| 1 | 0.007 | 0.3 | 6.3 | 11±4.7 | Z1 | ≈ 11±4.7 |
| | | 0.68 | 14.0 | | Convex | |
| | 0.307 | 0.62 | 12.5 | 17±4.5 | Z2/AdZ/Z2 | ≈ 6±1.6 |
| | | 0.94 | 21.4 | | Cv/Plan/Cv | |
| 2 | 0.383 | 0.43 | 11.9 | 13.5±1.6 | AdZ | ≈ 4.8±0.6 |
| | | 0.94 | 13.7 | | Planar | |
| | 0.554 | 0.27 | 8.3 | 11.3±3 | Z1 | ≈ 11.3±3 |
| | | 0.57 | 13.7 | | Convex | |
| | 0.675 | 0 | 4.2 | 9±4.8 | Z1 | ≈ 9±4.8 |
| | | 0.5 | 13.1 | | Convex | |

Table 1. Alteration in axial temperature gradient measured in empty crucible during the two runs of $Ca_{1-x}Sr_xF_2$ boules with different content. The assessed indirectly real *T*-gradient ahead the CF, $G_{real}$, is also given.

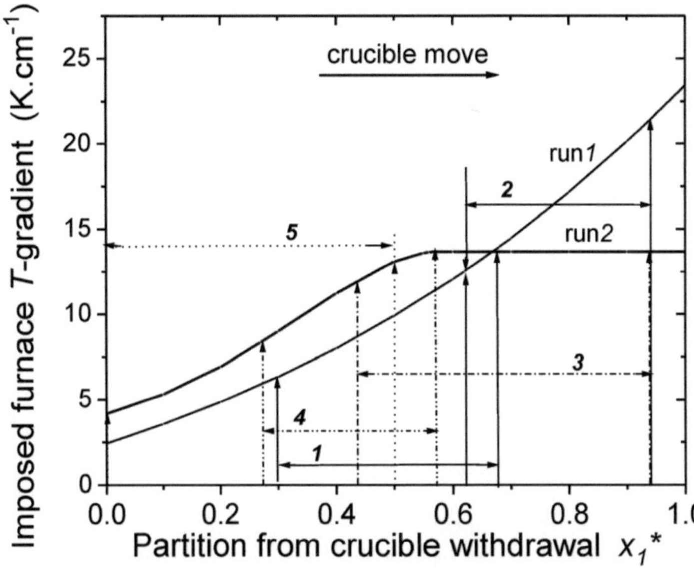


Fig. 7. Axial temperature gradient, measured in non-loaded multicameral crucible, as a function of crucible position during the two growing runs of $Ca_{1-x}Sr_xF_2$ boules with different content. Run*1*: 1 – $x = 0.007$ and 2 – $x = 0.307$; Run2: 3 – $x = 0.383$, 4 – $x = 0.554$ and 5 – $x = 0.675$.

Despite these common arguments assist the conducted analysis, its obligatory preciseness demands, however, both Sr to Ca interdiffusion coefficient and stability function dependences on the ratio of $G_{real}$ to $V_{real}$ being obtained by using corresponding fitting equations for $G_{real} = f_G(x_1^*)$ and $V_{real} = f_V(x_1^*)$. The first relationship follows directly from $G_{fur} = f_G'(x_1^*)$ shown in Fig. 7 by relevant reduction of 2.8 depending on whether the CF sifts inside or outside the LR (AdZ).

### 7.2.6 Determination of the real crystallization rate

The other factor, influencing the interdiffusion coefficient and, respectively the magnitude of stability function, is the real crystallization rate $V_{real}$. As discussed elsewhere (Mouhovski et al., 2011), this factor may differ - sometimes significantly - from the set SC, $V_{cru}$ in conjunction with the altering thermal conditions into the load where the shift in CF-positions at constant furnace supply should determine corresponding divergences of $V_{real}$ from $V_{cru}$. Nevertheless, as illustrates Fig. 6, $V_{real}^*$, vs. $x_1^*$ dependence should stand one and the same for all boules since they have been grown in practical at equal conditions where one the same relative CF-shift occurs in all cameras at any crucible cross-section. As seen for both runs in study (Fig. 8) the real CR remains less than that of the SC up to the moment when the cameras' tips cross-section passes into Z2.

Specifically for run*1* the surge of $V_{real}^*$ up to 25% above 1 can be explained with abrupt enlargement for the total heat flux towards Z2 caused by a sharp decrease of the relevant effective thermal resistance, which constitutional radial part just disappears when all the 9 rings move at positions in Z2 below the lower boundary cross section of LR.

In case no rings is being fixed on crucible stem (run 2), an extra-heat release in radial direction starts significantly earlier, when the crucible bottom plane approaches the lower boundary section of the AdZ. Then the further crucible withdrawal is peculiar in approaching approximately constant acceleration of the $V_{real}^*$ (Fig. 8) Such behavior is a consequence of the rise of 5 K/h imposed on the upper heater temperature $T_1$ (Fig. 1) whereby additional heat is supplied and flows into the inside of the upper load section. This way the heat losses resulting from the move into Z2 by a gradually enlarging lateral surface of the crucible are compensated with a surplus that causes a relevant downward shift of the CF corresponding to the observed constant increase of $V_{real}^*$. The smooth transition of a constant rate below the SC to a rate slightly in excess of SC is a ground for normal growth of boules having high optical quality.

The studied dependence manifests completely different behavior in case a gradual rise of 8.6 K/h is being imposed to lower heater temperature $T_2$ (Fig. 1, run*1*) for compensating the heat losses occurring through the movement into Z2. This manner it turns out not possible to adjust $V_{real}$ * to the favorable region around unity; the chosen heat increase is rather high leading to such fast lowering for CF-position that push $V_{real}^*$ to decline abruptly to values far below 1 and even for a while it stays below 0 (Fig. 8, run*1*). As a result of this procedure the $IF_{L/S}$-stability as well as the normal growth would both be severely disturbed due to the development of a strongly concave shape to the CF causing increased impurity incorporation and more pronounced supercooling effect. Besides a small portion nearby the surface of the already grown boules melts again when $V_{real}^*$ drops below zero, a phenomenon which would create additional growth anomalies and failure in crystal structure. Hence one may anticipate the top section of some run*1*-boules to possess significantly worse values for the main optical characteristics as compared to those ones of the lower boules' sections.

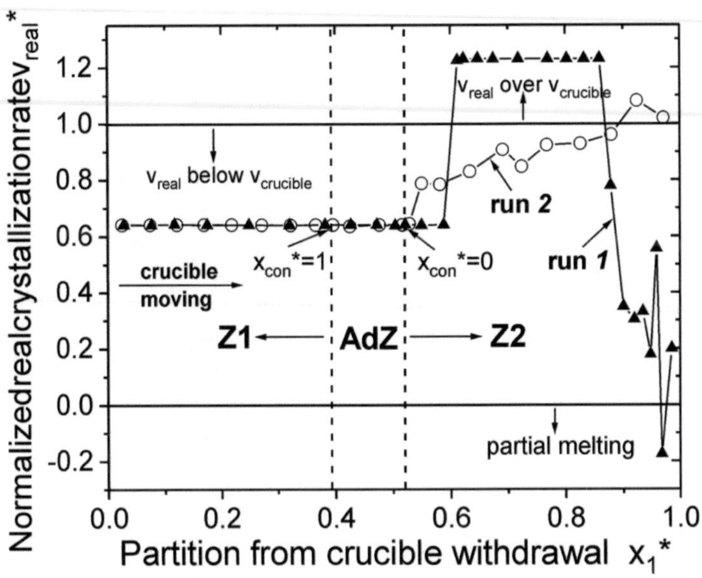


Fig. 8. Relative alteration of the real crystallization rate during two growing runs carried out under different thermal conditions for a series of $Ca_{1-x}\,Sr_xF_2$ boules.

### 7.2.7 Determination of interdiffusion coefficient

Calcium to strontium interdiffusion coefficient $D_{Ca/Sr}$ – according to formula (11) – depends primary on the critical value of stability function $F(x)_{crit} = m.\Delta x$ that has to be determined from the compositional phase diagram for studied solid solution system (Fig. 2). Besides, being proportional to this purely substantial thermodynamic factor, $D_{Ca/Sr}$ is a linear function of the ratio $V_{real}/G_{real}$ of the real CR to real *T*-gradient established ahead the moving CF. Despite this ration seems being depended solely on the imposed growing conditions by the set SC and longitudinal *T*-gradient in the furnace, both – the real CR and the real *T*-gradient – appear functions of CF-position in each particular camera, which in turn, depends on the composition of the loaded portions/grown boules. At this junction the chosen variations in apparatus factors for the two runs cause considerable differences between corresponding values for $D_{Ca/Sr}$ (Fig. 9). Hence, in the cases under consideration to take an average value for this important mass-transport parameter in stability function formula and relevant CMSc criteria may lead to incorrect assessment about the growth mechanism expected.

Indeed, the presented in Fig. 9 five representative curves for the two runs differs substantially each other. Such result means the product of critical stability function $F(x)_{crit}$ and $V_{real}/G_{real}$-ratio affects in complex way and mostly considerably the interdiffusion for the two types alkali earth cations into building in fluorite lattice. Specifically two order of magnitude differences are found out at $x = 0.6$ for run*1*/run*2* ratio. Nevertheless the run*1* dataset within $0.62 < x < 0.94$ interval show a trend for fast lowering $D_{Ca/Sr}$ owing to a combined effect of decreasing $F(x)_{crit}$ and increasing $G_{real}$ while the alteration of $V_{real}$ stays with less importance. Again relatively large values for $D_{Ca/Sr}$ ($\approx 4\times10^{-4}$ cm$^2$ s$^{-1}$) are found for run2 at $x$ within 0.62–0.86 that is due, evidently, to established low but constant real $T$-gradient. Here the abrupt lowering of $V_{real}$ starts to influence on $D_{Ca/Sr}$ only at $x > 0.86$. It seems the slowest, nearly constant interdiffusion to occur under run2-conditions in the boule with a largest Sr-content (x = 0.675) since then the $G_{real}$ is equal approximately to set furnace $T$-gradient, $G_{fur}$, while the $V_{real}$ to $V_{crucible}$ ratio is being maintained uniformly below 1 (0.64) during the whole run.

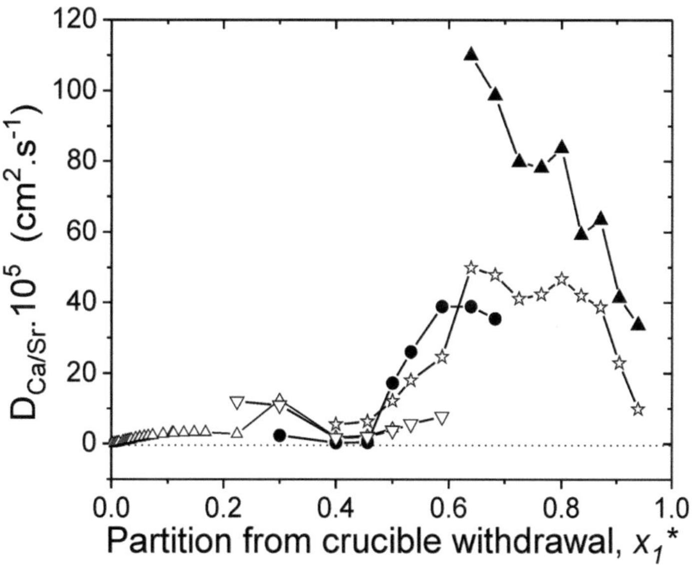


Fig. 9. Interdiffusion coefficient in growing of $Ca_{1-x}Sr_xF_2$ solid solution system as a function of the partition from total crucible withdrawal, $x_1$* at different strontium content $x$ in the simultaneously grown boules under particularly chosen thermal conditions (see Fig. 1). Run 1: (●) – $x$ = 0.007; (▲) – $x$ = 0.307. Run 2: (★) – $x$ = 0.383; (▽) – 0.554; (△) – 0.675.

The performed analysis of the compositional dependence for interdiffusion coefficient manifests how important is to be known (determined precisely or assessed correctly) all factors that influence this mass-transport coefficient so that the stability criterion to outline correctly the region of normal growth at minimum supercooling effect that may ensure simultaneous growing of boules - solid solutions of $CaF_2$–$SrF_2$ mixed system - with different composition.

### 7.3 Stability criterion and crystal quality

The unifying presentation of critical stability function vs. the partition from crucible withdrawal during the two growing runs (Fig. 10) appears especially convenient for comparative analysis. The calculated curve manifests certain similarity with the curve in Fig. 4 that, however, should not mislead the reader because the variables are completely different. Smoothing the curve, its extreme minimum of only ≈ 0.5 K appears within 0.4–0.46 that determines position of the crucible just before its middle cross-section. Around this minimum the boules should grow at very low melt supercooling but its exact position as regards boules' height is compositionally dependent via really established local $T$-gradient and CR. All area below the curve outlines growing conditions with unacceptable level for melt supercooling.

Specifically at *run1* conditions the *F*-function goes down steeply during the growth of approximately the first ⅓ of the height of the boules with utmost low Sr-content ($x \leq 0.1$) as nucleation starts at relatively high $F(x_1^* = 0.3) = 3.7$ K ( $x = 0.007$).

In accordance with curve's course the nucleation/growth conditions change dramatically for boules with higher $x$ approaching the limiting value of 0.307 (run*1*) whereat the entire process will proceed at significant melt supercooling (3.5 K ≤ $F(x_1^*)$ ≤ 3.95 K).

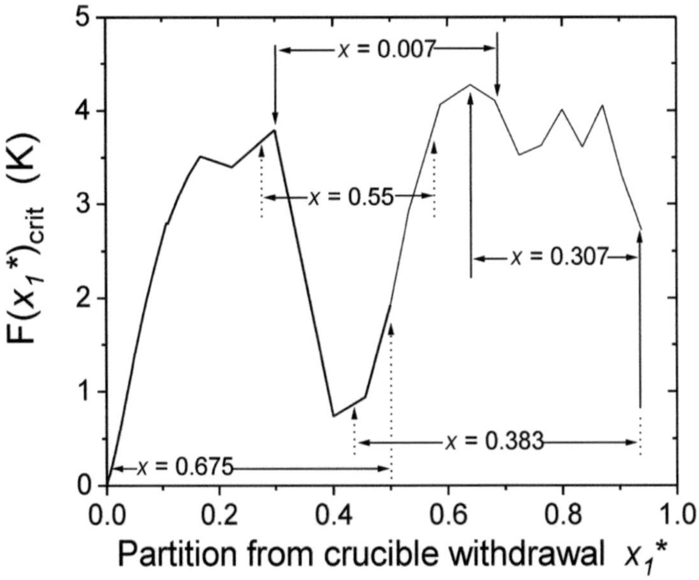


Fig. 10. Critical stability function $F(x_1^*)_{x=const}$ for calcium strontium fluoride solid solution system as a function of crucible movement towards Z2 for two specific run's growing conditions and parameter - strontium content into the grown boules. The limiting strontium concentrations for the two runs are marked by arrows: run*1* - 0.007–0.307; run2 - 0.383–0.675. $x = 0.55$ corresponds to boule with the lowest light extinction measured within UV - NIR spectral range.

On the other hand the growth of both boules, referring to limiting Sr-content $x$ of 0.383 and 0.675 (run2), should start at very low values for $F(x_1^*)$ but the function increases rapidly thereafter. Such behaviour suggests for fast deteriorating growing conditions. Hence the crystal quality for run2-boules can expect being worse as compared to that of run*1*-boules. Certain correlation is found analysing comparatively the light-extinction spectra obtained within the two series of windows between ser.1 (prepared from the lowest cylindrical boules' section and ser.2 (prepared from higher disposed section, distant at least to 0.6 cm from the first one. Nevertheless more precise analysis is being implemented using a new parameter, defined as the difference in stability function values determined at upper and lower boundary plane sections for given series windows, $\Delta F(x_1^*)$ that is being juxtaposed for particular boule to relevant discrete values of $E_\lambda$-spectra within UV, Vis and NIR (Fig. 11).

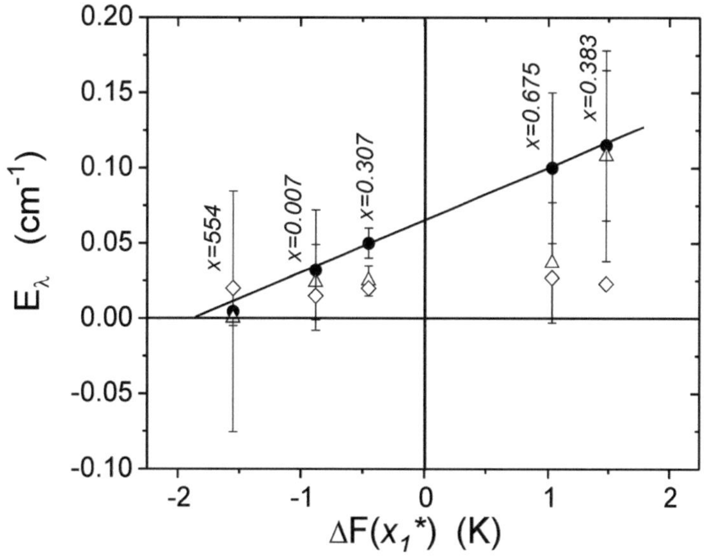


Fig. 11. Average light extinction for three λ within UV (●), Vis (△) and NIR (◇) spectral regions measured in ser.1-optical windows, finished from the lowest cylindrical section of the five representative boules vs. the difference in stability function values at coordinates corresponding respectively to upper and lower windows plane sections.

As seen, significant variations for $E_\lambda$ are established along the thickness of only 0.6 mm for ser.1-windows. Besides, the found deviations differ in conjunction with spectral region considered. Interestingly a linier dependence is fitted for dataset obtained into UV region that predetermines certain local structural inhomogeneities along the whole height of the grown boules. To check this suggestion all 17 grown boules are testified as regards possible longitudinal alteration of their optical transmission measured in corresponding pairs of windows. Relevantly calculated differences between coefficients of light extinction, ($E_{\lambda(win2)}$ – $E_{\lambda(win1)}$), are followed as a function of corresponding average quantities $E_{\lambda(win1+win2)/2)}$ for each particular boule.

The data points obtained (Fig. 12) are found to scatter significantly within interval of less than 0.1 cm$^{-1}$ as mostly differences (≈ 65%) appear with positive sign that stands for gradually degenerating optical quality towards the tip of corresponding boules. It is seen as well the deviations of positive values from zero line to scatter more broadly as comparing those for the boules where the sign of ($E_{\lambda(win2)}$ – $E_{\lambda(win1)}$) stays negative.

The linear fit reveals a very strong correlation in the VIS and IR with R > 0.94 that becomes significantly weaker in the UV (R ≈ 0.65) [*100*]. The later indicates a higher sensitivity to crystal imperfection from any variations in growth conditions, manifested by relevant alteration of the stability function, at the shorter wavelengths where even the smallest structural defects influence on the regularity of the crystal lattice. This result is in accordance with the followings in Fig. 10 where the light extinction just in the UV region depends essentially linearly on the newly introduced variable $\Delta F(x_1^*)$, that gives the direction of the relative $F(x_1^*)$ alterations. Hence the development of supercooling effect – by its amplifying or attenuating – gains a decisive importance upon the regularity of crystal lattice formation.

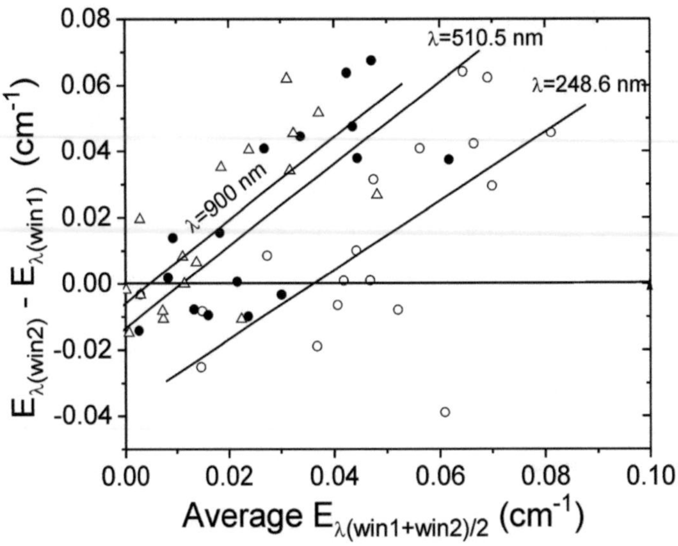


Fig. 12. Difference in light extinction coefficient obtained for particular wavelength within UV–IR in pairs of windows prepared from non-adjacent parallel sections of the grown boules with different content $x$ vs. average light extinction for corresponding pairs of windows at λ: 248.6 nm (○), 510.6 nm (•), and 900 nm/6.45 μm (△).

### 7.4 Normal growth criterion

The criterion for normal growth is being applied in order to assess how effectively the latent heat of fusion $L_{molt}$ can modify the growth mechanism for studied solid solution fluoride

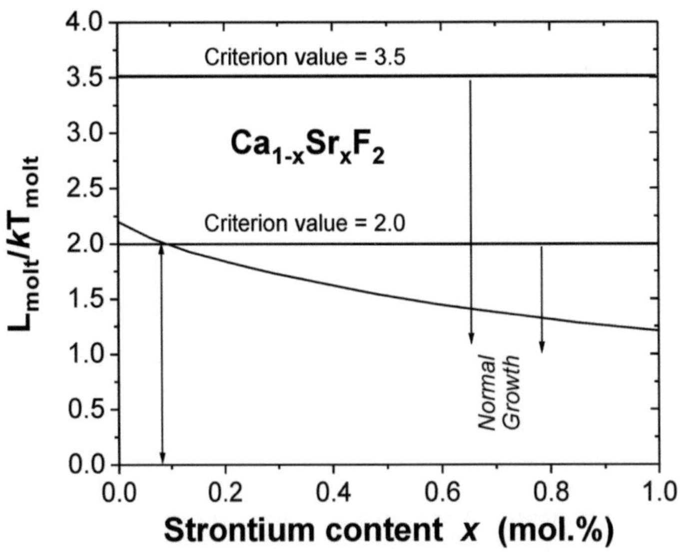


Fig. 13. Compositional dependence of normalised latent heat of fusion for studied $Ca_{1-x}Sr_xF_2$ solid solution system. Two cited in the literature boundary values for proceeding of normal growth are denoted correspondingly.

system, which involves single normal isotropic growth to a strongly anisotropic one with a resulting laminar distribution of the second (Sr) component. It is seen (Fig. 13) the studied $L_{molt}/kT$ compositional dependence crosses the criterion lower limiting value of 2.0 at $x \approx 0.08$ mole% whereas the curve is positioned entirely below the criterion upper limiting value of 3.5. At this junction one may suppose only two of run*1*-boules, the composition of which satisfies $x \leq 0.08$ mole% condition, being grown, eventually, under cellular, strongly anisotropic growth mechanism. However the results from the optical measurements and several microscopic observations (Mouchovski et al., 2011) do not suggest such mechanism to initiate dendroidal/cellular sub-structure in any of investigated sections (optical windows) for studied boules. Hence, taking criterion upper limit of 3.5, it may be concluded the growing conditions for both runs were ensured proceeding of normal growth in the used multicameral crucibles.

## 8. Conclusion

Simple constitutional melt supercooling criterion for liquid to solid interface stability may be applied for reliable assessment of the most favourable growing conditions established in batch growth, by optimised BS technique, of $Ca_{1-x}Sr_xF_2$ boules with broadly varying composition. The stability function representing the supercooling effect reveals specific compositional dependence, following the alterations arisen in $CaF_2$–$SrF_2$ phase diagram when for $CaF_2$ is being used fluorspar with different m.p. than that of the chemical reactive. The critical values for stability function vary along the boules in accordance with their composition, determining different positions wherefrom the crystallization starts to propagate. These interfering phenomena are grounded on specificity of the thermal field, established into loaded crucible, which can be controlled effectively by original appliance in growing apparatus for regulating the heat exchange into the furnace unit.

It is possible the axial *T*-gradient, measured into the furnace, being reduced in accordance with the positioning of the crucible into the thermal field established, to be recognized as the real *T*-gradient in stability function presentation.

The differences in stability function increasing or lowering values along definite sections of the boules, being linearly dependent on the relevant changes - positive or negative, respectively - in the key optical characteristics, bring significant valuable information about expected quality of mixed fluoride crystals with different composition, grown simultaneously at varying growing conditions. At that, the real crystallization rate can differ significantly on the set speed of crucible, thus imposing a need for supplementary correction in stability function magnitude along the boules' height.

Both, the alteration in stability function together with changeable ratio of real crystallization rate to real *T*-gradient, lead to great differences for interdiffusion behaviour in each section of the grown boules. This result should draw the researchers attention not to take automatically the view for constant interdiffusion processes during the growth of such complex solid solution fluoride systems but to provide supplementary investigation especially in cases of simultaneous growth of boules with different compositions.

The exact position of equilibrium segregation coefficient and interdiffusion coefficient into compositional space is of substantial importance for correct determination of stability function magnitude and application of corresponding supercooling criterion. The deviation of equilibrium segregation coefficient from unity may attain 20% at compositions with small or large $x$ that suggests considerable re-distribution in cation sub-lattice, predetermining stoichiometric changes along the boules' height.

The simple supercooling criterion manifests as well the importance for removing of any oxygen-containing contaminants from the crystallization zone (CZ). In case fluorspar is being used as starting material for growing of single or mixed fluoride compounds, it may be processed on original high-$T$ procedure, involving some scavenger's reactions in high vacuum, for substantial reduction - up to trace concentrations - of mostly residual metal oxides, product of decomposition of the accessory minerals, as well as deeply adsorbed on fluorspar grains oxygen anions. Possible penetration of ionised oxygen and water vapour molecules from vacuum ambient to melt bulk can be suppressed considerably when crucible construction provides Knudsen diffusion to dominate into the channels, relating the free spaces inside the crucible.

According to the simple supercooling criterion all impurities with effective segregation coefficient remaining sufficiently close to 1 (as appears all RE elements) and level of concentration within ppm-range do not cause any supercooling effect of significance.

The followings from applying the simple supercooling criterion should be apprehended cautiously since the high transparency for growing solid solution fluoride crystals and the semi-transparency of their melts within relatively large spectral range may change radically the thermal fluxes throughout the load leading to supplementary supercooling ahead the CF, the effect of which should be subject to analysis.

## 9. Acknowledgements

The author takes this opportunity to express special thanks to Prof. Tzvety Tzvetkov DSc CEO BG H2 Society who assisted in writing the survey and funded for its publishing. The author thanks Svilen Genchev for assistance in performing part of the experimental.

## 10. References

Alfintzev, G.A. & Ovsienko, D.E. (1980). Peculiarities of melt growing crystals of substances with various entropy of fusion. *Rost Kristallov* (Nauka, Moskwa), Vol.13, 121–123 (in Russian), ISBN/ISSN: 0485-4802.

Chalmers B. (1968). *Principles of solidification*, John Wiley & Sons, New York.

Chern'evskaya, E.G. & Anan'eva, G.V. (1966). Structure of $CaF_2$-, $SrF_2$-, and $BaF_2$-Based Mixed Crystals, *Fizika Tverdogo Tela* (Leningrad), Vol.8, No.1, 216–219 (in Russian), ISSN 1063-7834.

Chernov, A.A. (1984). *Modern Crystallography III. Crystal Growth* (Springer Ser. Solid-State Sci.), Vol.36, Springer, Berlin, Heidelberg 1984).

Deshko, V.I.; Kalenichinko, S.G.; Karvatzkii, A.Y. & Sokolov, V.A. (1990). Stability of temperature conditions during vertical directed crystallization of calcium fluoride at radiative-conductive heat exchange. *Optico-mehanicheskaya promyshlennost*, No.1, 46–48 (in Russian), ISSN: 0030-4042.

Deshko, V.I.; Karvatzkii, A.Y.; Sokolov, V.A. & Hlebnikov, O.E. (1986). Temperature distribution in crystallization furnace at continuous growth of calcium fluoride crystals. *Optico-mehanicheskaya promyshlennost*, No.8, 28–31 (in Russian), ISSN: 0030-4042.

Djurinskij, B.F. & Bandurkin, T.A. (1979). Regularity in properties of lanthanides into inorganic materials, *Izvestiya Akademii Nauk SSSR, Neorganicheskie materialy*, Vol.15, 1024–1027 (in Russian), ISSN: 0002-337X.

Fedorov, P.P. (1991). Association of Point Defects in Non Stoichiometric $M_{1-x}$ $R_xF_{2+x}$ Fluorite-Type Solid Solutions, *Bulletin of Society Catalan Cien*, Vol. 12, No. 2, 349–381.

Fedorov, P.P. (2000). Heterovalent isomorphism and solid solutions with a variable number of ions in the unit cell. *Russian Journal of Inorganic Chemistry*, Vol. 45, No. 3, 268–291, ISSN PRINT: 0036-0236. ISSN ONLINE: 1531-8613.

Fedorov, P.P.; Buchinskaya, I.I.; Ivanovskaya, N.A. et al. (2005). $CaF_2$-$BaF_2$ Phase Diagram, *Doklady Akademii Nauk*, Vol. 401, No.5, 652–654 (in Russian), ISSN PRINT: 0869-5652.

Fedorov, P.I. & Fedorov, P.P. (1982). *Grounds of technology for production of super-pure substances*, MIHM, Moskwa (in Russian).

Fedorov, P.P.; Turkina, M.; Meleshina, V.A. & Sobolev, B.P. (1988). Conditions of formation of cellular sub-structure in mono-crystals solid solutions inorganic fluorides with fluorite structure. *Rost Kristallov*, (Nauka, Moskwa), Vol.17, 198–216 (in Russian), ISBN/ISSN: 0485-4802.

Ito, M.; Goutaudier, Ch.; Guyot, Y.; Lebbou, K.; Fukuda, T. & Boulon, G. (2004). Crystal Growth, $Yb^{3+}$ Spectroscopy, Concentration Quenching Analysis and Potentiality of Laser Emission in $Ca_{1-x}Yb_xF_{2+x}$, *Journal of Physics: Condensed Matter*, Vol.16, No. 8, 1501–1521, ISSN 0953-8984 (Print). ISSN 1361-648X (Online).

Jackson, A. (1958). Mechanism of growth, Liquid Metals and Solidification. *American Society for metals*, Cleveland, Ohio, 174–186, ISSN: 0096-7416.

Jackson, K.A. (2004). Constitutional Supercooling Surface Roughening. *Journal of Crystal Growth*, Vol.264, 519–529, ISSN: 0022-0248.

Kaminskii, A.A.; Rhee, H.; Eicher, H.J.; Bohaty, L.; Becher, P. & Takaichi, K. (2008). Wide-band Raman Stoks and Anti-Stoks lasing comb in a $BaF_2$ single crystal under picosecond pumping. *Laser Physics Letters*, Vol. 5, No.4, 304–310, ISSN print: 1612-2011, ISSN electronic: 1612-202X.

Kazanskii, S.A.; Ryskin, A.I.; Nikiforov, A.E.; Zaharov, A.Yu.; Ougrumov, M.Yu. & Shakurov, G.S. (2005). EPR Spectra and Crystal Field of Hexamer Rare-Earth Clusters in Fluorites, *Physics Review B*, Vol.72, 014127-1–014127-11, ISSN: 0556-2805.

Klimm, D.; Rabe, M.; Bertram, R.; Uecker, R. & Parthier, L. (2008). Phase diagram analysis and crystal growth of solid solutions $Ca_{1-x}Sr_xF_2$, *Journal of Crystal Growth*, Vol. 310, No.1, 152–155, ISSN: 0022-0248.

Kuznetsov V.A. & Fedorov P.P. (2008). Morphological Stability of Solid-Liquid Interface during Melt Crystallization of $M_{1-x}R_xF_{2+x}$ Solid Solutions. *Inorganic Materials*, Vol. 44, No.13, 1434-1458, ISSN PRINT: 0020-1685; ISSN ONLINE: 1608-3172. DOI: 10.1134/S0020168508130037

Lucca, A., Debourg, G.; Jacquemet, M.; Druon, F.; Balembois, F.; Georges, P.; Camy, P.; Doualan, J.L. & Moncorge, R. (2004). High-Power Diode-Pumped $Yb^{3+}$:$CaF_2$ Femtosecond Laser, *Optics Letters*, Vol.29, No.23, 2767–2769, ISSN:0146-9592 (print) ISSN: 1539-4794 (online).

Maier, J. (1995). Ionic Conduction in Space Charge Regions, *Progress in Solid State Chemistry*, Vol.23, No.3, 171–263, ISSN: 0079-6786.

Mouchovski, J.T. (2006). *Growing of Optical Crystals of Calcium Fluoride*. Scientific Work, Central Scientific Technical Library at National Center for Information and Documentation, Sofia, Bulgaria (in Bulgarian), ISSN: 1312-6164.

Mouchovski, J.T. (2007a). Control of oxygen contamination during the growth of optical calcium fluoride and calcium strontium fluoride crystals, *Progress in Crystal Growth and Characterization of Materials*, Vol.53, 79–116, ISSN: 0960-8974.

Mouchovski, J.T. (2007b). Growing of mixed crystal compounds based on calcium and strontium fluorides, *Bulgarian Chemical Communications*, Vol.39, No.1, 3–8, 2007, ISSN: 0324-1130.

Mouchovski, J.T.; Genov, V.B.; Pirgov, L. & Penev V.Tz. (1999). Preparation of $CaF_2$ precursors for laser grade crystal growth. *Materials Research Innovations*, No.3, 138–145, ISSN (printed): 1432-8917. ISSN (electronic): 1433-075X.

Mouchovski, J.T.; Penev, V.Tz. & Kuneva, R.B. (1996). Control of the growth optimum in producing high-quality $CaF_2$ crystals by an improved Bridgman-Stockbarger technique. *Crystal Research and Technology*, Vol.31, No.6, 727–737 Print-ISSN 0232-1300. Online-ISSN 1521-4079.

Mouhovski, J.T.; Temelkov, K.A. & Vuchkov, N.K. (2011). The Growth of Mixed Alkaline-Earth Fluorides for Laser Host Applications. *Progress in Crystal Growth and Characterization of Materials*, Vol. 53, 1–41, ISSN: 0960-8974.

Mouhovski, J.T.; Temelkov, K.A.; Vuchkov, N.K. & Sabotinov, N.V. (2009a). Calcium strontium fluoride crystals with different composition for UV-laser application: control of growing rate and optical properties, *Compes Rendus*, Vol.62, No.6, 687–694, ISSN: 1251-8050.

Mouhovski, J.T.; Temelkov, K.A.; Vuchkov, N.K. & Sabotinov, N.V. (2009b). Simultaneous growth of high quality $Ca_{1-x}Sr_xF_2$ boules by optimized Bridgman-Stockbarger apparatus. Reliability of high-transmission measurements, *Bulgarian Chemical Communications*, Vol.41, No.3, 253–260, ISSN: 0324-1130.

Nassau, K. (1961). Application of the Czochralski Method to Divalent Metal Fluorides. *Journal of Applied Physics*, Vol.32, No.10, 1820–1825, ISSN: 0021-8979.

Popov P.A.; Chernenok E.V.; Fedorov P.P.; Kuznetsov S.V.; Konyushkin V.A. & Basiev T.T. (2006). Heat Conduction of Single Crystals of Heterovalence Solid Solutions of Ytterbium and Praseodymium Fluorides in Calcium Fluoride, *Kondensirovannye Sredy Mezhfazovyh Granitz*, No.4, 320–321 (in Russian), ISSM: 160-6-867X.

Popov, P.A.; Fedorov, P.P.; Konyushkin, V.A.; Nakladov, A.N.; Kuznetsov, S.V.; Osiko, V.V. & Basiev T.T. (2008). Thermal Conductivity of Single Crystals of $Sr_{1-x}Yb_xF_{2+x}$ Solid Solution. *Doklady Physics*, Vol. 53, No. 8, 413–415 (in Russian), ISSN PRINT: 1028-3358. ISSN ONLINE: 1562-6903. DOI: 10.1134/S1028335808080016

Popov, P.A.; Fedorov, P.P.; Kuznetsov, S.V.; Konyushkin, V.A.; Osiko, V.V. & Basiev, T.T. (2008). Thermal Conductivity of Single Crystals of $Ca_{1-x}Yb_xF_{2+x}$ Solid Solutions, *Doklady Physics*, Vol. 53, No. 4, 198–200 (in Russian), ISSN PRINT: 1028-3358. ISSN ONLINE: 1562-6903. DOI: 10.1134/S102833580804006X

Popov, P.A.; Fedorov, P.P.; Kuznetsov, S.V.; Konyushkin, V.A.; Osiko, V.V. &
Basiev, T.T. (2008). Thermal Conductivity of Single Crystals of $Ba_{1-x}Yb_xF_{2+x}$ Solid Solution. *Doklady Physics*, Vol. 53, No. 7, 353–355 (in Russian), ISSN PRINT: 1028-3358. ISSN ONLINE: 1562-6903. DOI: 10.1134/S1028335808070045

Sekerka, R.F. (1965). A stability function for explicit evaluation of the Mullins-
Sekerka interface stability criterion. *Journal of Applied Physics*, Vol.36, No.1, 264–268, ISSN (printed): 0021-8979, ISSN (electronic): 1089-7550.

Sekerka, R.F. (1968). Morphological Stability, *Journal of Crystal Growth*, Vol.3/4, 71–81, ISSN: 0022-0248.

Sobolev, B.P. & Fedorov, P.P. (1978). Phase Diagrams of the $CaF_2$-(Y, Ln)$F_3$ Systems. I. Experimental. *Journal of the Less-common Metals*, Vol.60, 33–46, ISSN: 0022-5088.

Sobolev, B.P.; Golubev, A.M. & Herrero, P. (2003). Fluorite $M_{1-x}R_xF_{2+x}$ phases (*M* = Ca, Sr, Ba; *R* = Rare Earth Elements) as nanostructured materials. (2003). *Crystallography Reports*, Vol.48, No.1, 141–161, ISSN PRINT: 1063-7745.

Sorokin, N.I.; Buchinskaya, I.I.; Fedorov, P.P. & Sobolev, B.P. (2008). Electrical conductivity of a $CaF_2$-$BaF_2$ nanocomposite, *Inorganic materials*, Vol.44, No.2, 189–192, ISSN: 0020-1685.

Tiller W.A.; Jackson, K.A.; Rutter, J.W. & Chalmers. B. (1953). The redistribution of solute atoms during the solidification of metals. *Acta Metallurgica*, Vol.1, 428–437, ISSN: 0001-6160.

Turkina, T.M.; Fedorov, P.P. & Sobolev, B.P. (1986). Stability of the flat crystallization front during growing from melt of mono-crystals of $M_{1-x}R_xF_{2+x}$ solid solutions (where M = Ca, Sr, Ba; R - rare earth elements). *Kristallografiya*, Vol.31, No.1, 146–152 (in Russian), ISSN PRINT: 0023-4761.

Van-der-Vaal's, I.D. & Konstamm, F. (1936). Course in thermostatic, Vol.1, ONTI, Moskwa (in Russian).

Weller, P.F.; Axe, J.D. & Pettit, G.D. (1965). Chemical and Optical Studies of Samarium Doped $CaF_2$ Type Single Crystals, *Journal of Electrochemical Society*, Vol.112, 74–77, ISSN: 1945-7111 online; ISSN: 0013-4651 print.

Wrubel, G.P.; Hubbard, B.E.; Agladge, N.I. et al. (2006). Glasslike Two-Level Systems in Minimally Disordered Mixed Crystals, *Physical Review Letters*, 2006, Vol.96, 235503-235507, ISSN: 0031-9007.

Yushkin, N.P.; Volkova, N.V. & Markova, G.A. (1983). *Optical fluorite*, Nauka, Moskva (in Russian).

Zhigarnovskii, B.M. & Ippolitov, E.G. (1969). Phase Diagram of the $CaF_2$–$BaF_2$ System, *Izvestii Akademii Nauk SSSR, Neorganicheskie Materialy*, Vol.5, No.9, 1558–1562 (in Russian), ISSN: 0002-337X.

6

# Formation of Dissipative Structures During Crystallization of Supercooled Melts

Leonid Tarabaev and Vladimir Esin*
*Institute of Metal Physics, Ural Division of the Russian Academy of Sciences*
*Russia*

## 1. Introduction

The process of crystallization in a system far from equilibrium has features, which manifest themselves in the morphology, crystal growth velocity, and segregation of dissolved alloy components. So under conditions of high cooling rates of melt ($R \sim 10^6$ K/s), when the deep supercoolings are reached an irregular morphology of solidification, nonequilibrium «trapping» of impurity, and coexistence of crystalline and amorphous phases observed (Miroshnichenko, 1982). For sufficiently high growth velocities, i.e. for certain critical undercooling the sharp transition to a partitionless regime of crystallization will take place (Nikonova & Temkin, 1966). It afterwards was called as kinetic phase transition (Chernov & Lewis, 1967; Chernov, 1980; Temkin, 1970). The critical supercooling can reach large values: so for Ni-B alloy of ~ 200 ÷ 300 K (Eckler at al., 1992). The dissipative structures formed in such system, essentially influence on set of main properties of prepared material.

In the continuous growth model the boundary conditions for solute partitioning at the crystal-melt interface are established (Aziz & Kaplan, 1988; Aziz, 1994; Kittl at al., 2000). These conditions are used in models to explaining the experimental data on solute trapping, in particular, in the phase-field models (Ahmad at al., 1998; Ramirez at al., 2004; Wheeler at al., 1993). The dependences of the kinetic coefficient and the diffusion rate at the interface on the temperature are not usually considered. In models of dendritic growth used for the computation of the rapid solidification kinetics (Eckler at al., 1992, 1994) the diffusion rate and the kinetic coefficient in a collision-limited form are entered as independent fitting parameters. The method of computer simulation (Tarabaev at al., 1991a) allows study the formation of a complex morphology of the solid - liquid interface and it dynamics during a crystallization of pure metals (Tarabaev at al., 1991b) and metal alloys (Tarabaev & Esin, 2000, 2001). In this work the crystallization from one centre of a binary essentially nonequilibrium system is investigated in computer model (Tarabaev & Esin, 2007) that takes into account the temperature dependence of the diffusion coefficient and the nonequilibrium partition of dissolved component of the alloy (Aziz & Kaplan, 1988).

* Corresponding Author

## 2. Computer model

### 2.1 Kinetics of crystallization

The computer model is based on a finite difference method. The two–dimensional finite-difference grid divides the system into cells. Each cell is characterized by a volume fraction of a solid phase $g_S$. Assuming the normal mechanism of crystal growth the velocity of an interface motion $V$ in a two-phase cell ($0 < g_S < 1$) can be written as follows:

$$V = \beta \Delta T \, , \tag{1}$$

where $\beta$ - is anisotropic kinetic coefficient, $\Delta T$ - is kinetic supercooling at the interface:

$$\Delta T = T_E - T_I = T_M(1 - d_0 \kappa) - mc_I - T_I \tag{2}$$

Here $T_E$ - is equilibrium temperature, $T_M$ - is temperature of melting of first component, $d_0$ - is capillary length:

$$d_0 = \gamma_{SL}/Q \, , \tag{3}$$

Here $\gamma_{SL}$ - is surface tension of crystal – melt interface, $Q$ - is heat of melting, $\kappa$ - is interface curvature, $m$ - is the slope of the equilibrium liquidus line (without sign), $T_I$ and $c_I$ - are the temperature and the concentration in the liquid at the interface, respectively.

The nonequilibrium effect of solute partition at interface is described by expression (Aziz & Kaplan, 1988) for partition coefficient $k$:

$$k(V) = \frac{c_S}{c_L} = \frac{V/V_D + k_e}{V/V_D + 1 - (1 - k_e)c_L} \, , \tag{4}$$

where $c_S$ and $c_L$ - are concentration in the solid and in the liquid at the interface, respectively,

$$k_e = k_0 \, / \, k_0^A \, , \tag{5}$$

Here $k_0$ and $k_0^A$ - are the equilibrium partition coefficients of solute and solvent, respectively; $V_D$ - is the rate of diffusion:

$$V_D \equiv f \nu a \exp(-E_a/RT) = D \, / \, a \, , \tag{6}$$

Here $f$ - is geometric factor, $\upsilon$ - is the atomic vibration frequency, $a$ - is interatomic spacing, $E_a$ - is the activation barrier for diffusion through the interface, $D$ - is coefficient of diffusion at the interface. The partition coefficient depends on the ratio of velocity of crystallization $V$ to rate of diffusion $V_D$. Rate of diffusion is the ratio of the diffusion coefficient at interface to the interatomic spacing.

The kinetic effect includes both temperature and orientation dependences of the kinetic coefficient, whose polar diagram has the four-fold symmetry and the directions of the maxima $\beta$ coincide with the principal grid directions. Then the velocity of an interface motion $V$ can be written as (Chernov, 1980; Miroshnichenko, 1982):

$$V = \beta \Delta T = f' \nu a \exp(-E'_a/RT) Q \Delta T / R T_E T \, , \tag{7}$$

where $\beta$ - is the kinetic coefficient:

$$\beta = f'\nu a \exp\left(-E'_a/RT\right)Q/RT_E T \ , \tag{8}$$

$f'$ - is a factor of anisotropy, $E_a^{'}$ - is the activation barrier for atomic kinetics. The ratio of the interface velocity to the diffusion rate in (4) can be written as follows

$$\frac{V}{V_D} = \frac{f'}{f}\exp\left(-\frac{E_a^{'} - E_a}{RT}\right)\frac{Q\Delta T}{RT_E T}\ , \tag{9}$$

In the case when the energy of activation for atomic kinetics $E_a^{'}$ is equal to the energy of activation for diffusion $E_a$ at the interface from expression (9) follows that

$$\frac{V}{V_D} = \frac{f'}{f}\left(\frac{Q\Delta T}{RT_E T}\right). \tag{10}$$

The ratio $V/V_D$ characterizes of the deviation degree from equilibrium of the interface for given temperature ($\Delta T/T$) and entropy of melting ($Q/RT_E$). That is in terms of the atomic kinetics it signifies the ratio of a resulting flux of atoms to an exchange (equilibrium) flux at the interface. And the ratio $(f'/f)$ characterizes the degree of the anisotropy of crystal growth rate. In case of $V = V_D$ from the equation (10) the expression for a supercooling follows:

$$\Delta T^* = \frac{T_E}{1 + \left(f^{'}/f\right)\left(Q/RT_E\right)}\ , \tag{11}$$

which can be the criterion of transition to nonequilibrium trapping of dissolved component of the alloy at interface at supercoolings larger than this value. Graphic presentation of the equations (1, 4, 6, and 7) for the system *Fe-B* is shown in Fig. 1.

## 2.2 Heat - and mass transfer in a system

For each volume element $\Omega$ of a system from conservation laws follow the equations for fields of $c_L$, $c_S$, and $T$:

$$\frac{\partial c_L}{\partial t} = \frac{(1-k)c_L}{g_L}\frac{\partial g_S}{\partial t} + \frac{1}{g_L \Omega}\oint_S \left(g_L D_L(T)\vec{\nabla} c_L, d\vec{S}\right), \tag{12}$$

$$\frac{\partial (c_S g_S)}{\partial t} = k c_L \frac{\partial g_S}{\partial t}\ , \tag{13}$$

$$\frac{\partial T}{\partial t} = \alpha \nabla^2 T + \frac{Q}{C}\frac{\partial g_S}{\partial t}, \tag{14}$$

where $g_S$ and $g_L$ - are fractions of solid and liquid phases, $\alpha$ and $C$ – is the thermal diffusivity and capacity, respectively; the diffusion coefficient in the melt depends on the temperature:

$$D_L(T) = a^2 \nu \exp\left(-E_D/RT\right) \tag{15}$$

Here, we neglect diffusion in the solid phase and the thermal diffusivity $\alpha$ is accepted identical in both phases. The source in equation (14) (also in (12)) is simulated by algorithm developed in (Tarabaev at al., 1991a), using the expression for the change of a volume fraction of a solid phase in two-phase cell of a system:

$$\frac{\partial g_s}{\partial t} = \frac{V(\vec{n})l(\vec{n})}{\Omega}, \tag{16}$$

where $\vec{n}$ - is the local normal to the interface segment in a two-phase cell and $l\,(\vec{n})$ - is the area of the interface segment. The finite-difference scheme of the problem was formulated with regard for these equations, and the corresponding computer program was modified (Tarabaev & Esin, 2007).

We now use the dimensionless quantities:

$$\tilde{c} = (c_0/k_0 - c)/\Delta c_0, \quad \tilde{V} = V/v_0, \quad \tilde{V}_D = V_D/v_0, \quad \Delta\tilde{T} = \Delta T\, C/Q.$$

Here, $c_0$ - is the initial concentration of solute in melt,

$$\Delta c_0 = c_0\,(1 - k_0)/k_0\, v_0, \tag{17}$$

$$v_0 = \beta_0\, Q/C, \tag{18}$$

$\beta_0$ - is isotropic kinetic coefficient at the phase equilibrium temperature. The relation between the diffusion rate and the kinetic coefficient at the equilibrium temperature has the form:

$$V_D(T_E) = \varepsilon \beta_0 R T_E^2 / Q, \tag{19}$$

where $\varepsilon$ - is the factor which takes into account the difference between activation barriers for atomic kinetics and for the diffusion at the interface. We assume that

$$V_D(T_E) / v_0 = \varepsilon\, \Theta / (Q / RT_E) = 0.9, \tag{20}$$

here $\varepsilon = 0.31$ ($\varepsilon = 1$ at $E_a^{'} = E_a$), and

$$\Theta = (C / Q) T_E. \tag{21}$$

Dependence of growth velocity $\tilde{V}$ on supercooling of the melt $\Delta\tilde{T}$ is calculated in a maximum (+) for $(f'/f)_{\max} = 1.5$ and in minimum (-) for $(f'/f) = 1$ of orientation dependence of a kinetic coefficient, which also depends on the temperature.

Velocity of crystallization $\tilde{V}_0$ as function of supercooling in case of constant kinetic coefficient (without temperature dependence) is presented in Fig. 1. At the large supercoolings (dimensionless supercoolings are more approximately 0.2) the curve calculated from (7) essentially deviates from the linear growth law. Velocity $\tilde{V}$ as a function of $\Delta\tilde{T}$ has a maximum at the supercooling:

$$\Delta\tilde{T}_{V\max} = T_E C / Q \Big/ \left(1 + E_a^{'} / RT_E\right) \qquad (22)$$

We will note also, that velocity of crystallization becomes equal to the diffusion rate at values of supercooling $\Delta\tilde{T}^*$ which depend on the degree of anisotropy of kinetic coefficient that is on the ratio ( $f'/f$ ) in expression (11).

Values $\Delta\tilde{T}^*$ will be smaller at $\varepsilon < 1$. All these presented above effects are known separately, but in this model they are interdependent. Taking into account the diffusion as the limiting factor, the true $\tilde{V}(\Delta\tilde{T})$-curve will lie below than these dependences are calculated from (7) in the kinetic regime of crystal growth. The kind of a curve can be obtained as a result of the computer simulation. Crystal growth is controlled by a joint action of kinetic phenomena at the interface and heat transfer and second-component mass transfer in the system.

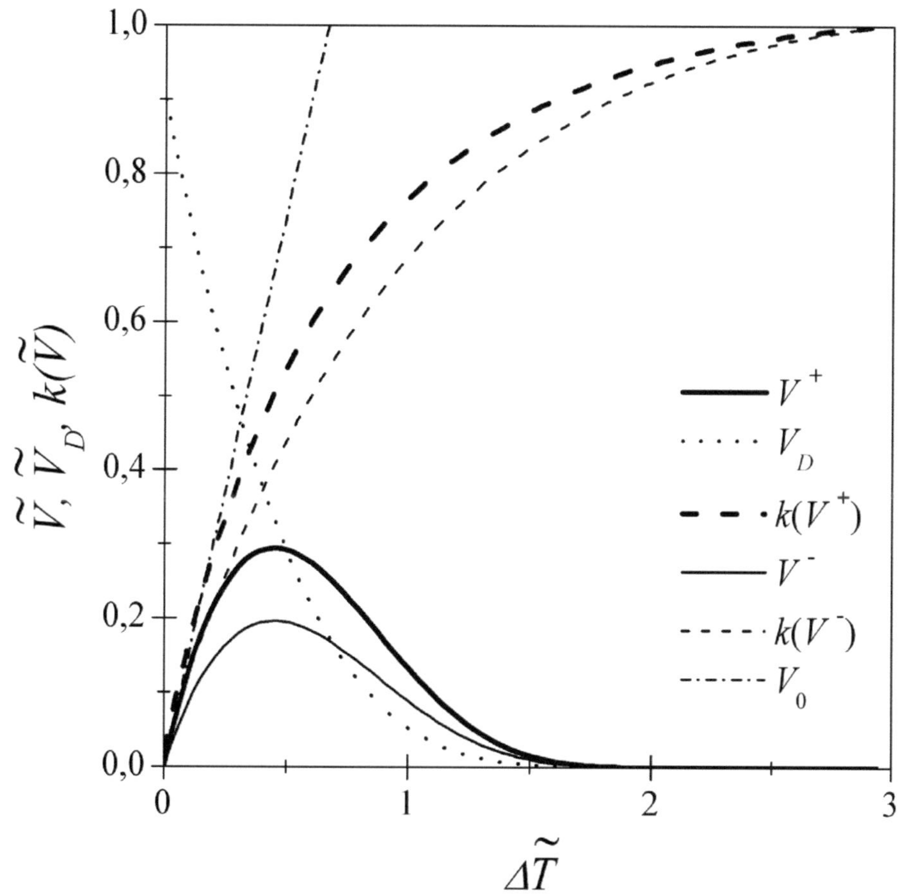


Fig. 1. Dependences of velocity of crystal growth $\tilde{V}$, diffusion rate $\tilde{V}_D$ and nonequilibrium partition coefficient $k$ ($V$) on supercooling $\Delta\tilde{T}$ of the $Fe$–$B$ melt with $\Theta = 2.96$, $E_a/RT_E = 5.55$, $Q/RT_E = 1.0$, and $k_0 = 0.015$. Curves drawn in maximum ($V^+$) and minimum ($V^-$) of the orientation dependence of kinetic coefficient (at $V^+/V^- = 1.5$).

## 3. Results of computer simulation

We consider the melt solidification from one centre in a two-dimensional system of different sizes (N $\Delta h$ x N $\Delta h$ cells). Here $\Delta h$ = 0.20 ( $\alpha$ / $v_0$) - is linear size of the unit cell.

The parameters of the problem approximately correspond to *Fe-B* system (Hansen & Anderko, 1957; Hain & Burig, 1983): the iron melting temperature is $T_M^{Fe}$ = 1809 K, the liquidus temperature is $T_E$ = 1803 K at an initial boron concentration $c_0$ = 0.04 wt % B ($\Theta = (C/Q)T_E$ = 2,96); $E_a$ = 83.8 kJ/mol ($E_a/RT_E$ = 5,55), $Q$ = 15.38 kJ/mol ($Q/RT_E \approx 1$), $Q/C$ = 609K, the equilibrium partition coefficient is $k_0$ = 0.015, and the slope of the liquidus line $m$ = -102 K/wt %. $D_L$ ($T_E$) = 5 × 10$^{-9}$ m$^2$/s, $\gamma_{SL}$ = 0.12 J/m$^2$, $\alpha$ = 0.7 × 10$^{-5}$ m$^2$/s. The characteristic scales of the problem ware as follows: the velocity $v_0 \approx 1.2 \times 10^2$ m/s, length ( $\alpha$ / $v_0$ ) ≈ 1 × 10$^{-7}$ m, and time ( $\alpha$ / $v_0^2$ ) ≈ 1 × 10$^{-9}$ s for $\beta_0$ = 0.2 m/(s K).

### 3.1 Morphology of dissipative structures formed during the solidification of a supercooled melt under conditions when the crystal growth is limited by diffusion of the dissolved component

We consider the crystal growth from of a single centre in a system of 1000 × 1000 cells (Fig. 2) and in a system of 500 × 500 cells (Fig. 3-5) with the following calculation parameters: time step $\Delta\tilde{t} = (v_0^2/\alpha)\Delta t$ = 0.02 and spatial step $\Delta\tilde{h} = (v_0/\alpha)\Delta h$ = 0.20. The computer simulation is realized in the systems with a fixed supercooling $\Delta T = \Delta T_{bath}$ with adiabatic boundary. The anisotropy of the orientational dependence of the kinetic coefficient $(f') = 6$. In these calculations (in this region of the melt supercooling) does not take into account the temperature dependence of the kinetic processes in interface crystal and diffusion of solute in the melt.

Morphology of dissipative structures formed during *Fe-B* melt crystallization and concentration fields in the system in some moments of the time $\tilde{t}$ for various values of bath

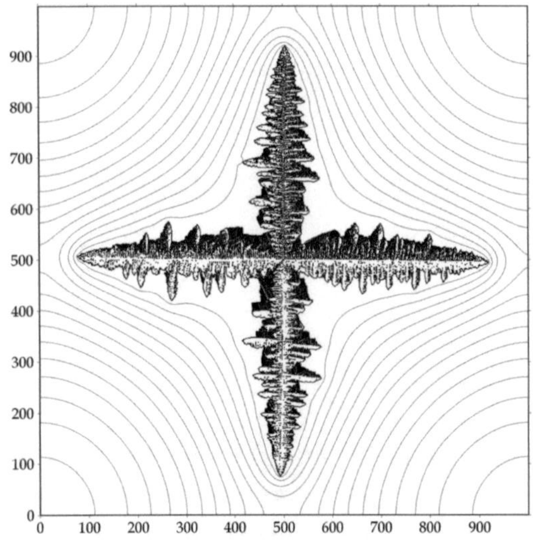

(a)

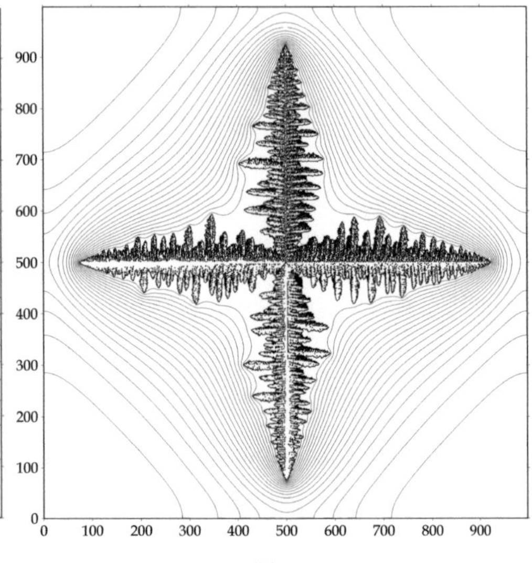

(b)

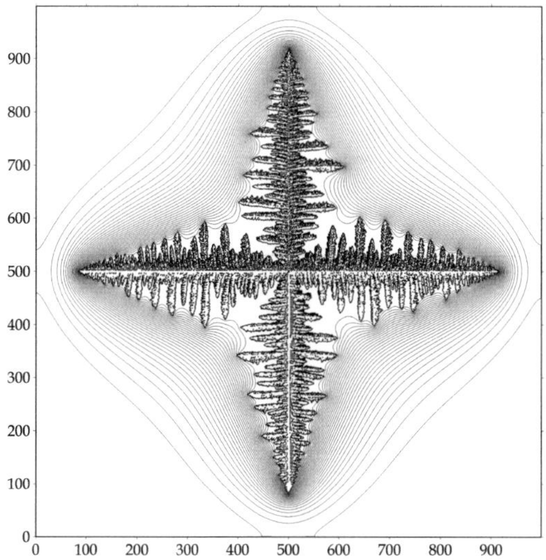


(c)

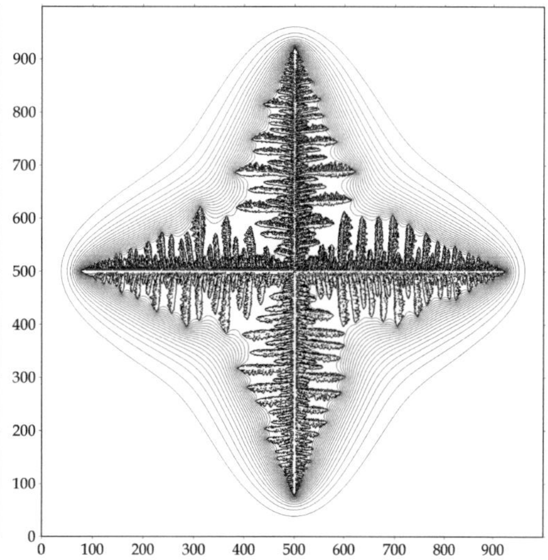


(d)

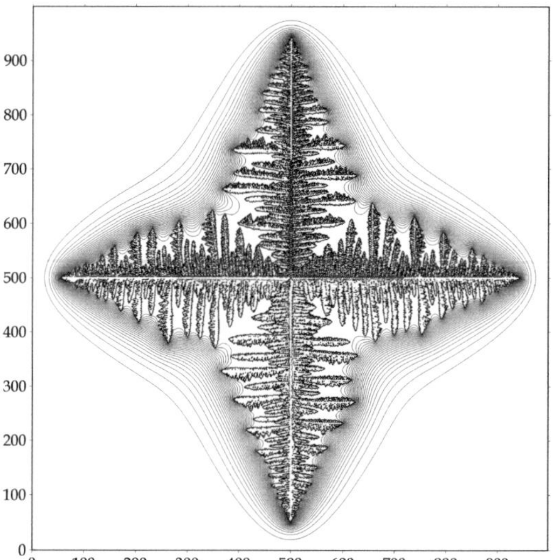


(e)

Fig. 2. Morphology of dissipative structures are formed during *Fe-B* melt crystallization and concentration fields (thin lines) in the system in some moments of the time $\tilde{t}$ for various values of bath supercooling $\Delta\tilde{T}_{bath}$ : (a) $\Delta\tilde{T}_{bath} = 0.01$, $\tilde{t} = 750$; (b) $\Delta\tilde{T}_{bath} = 0.02$, $\tilde{t} = 230$; (c) $\Delta\tilde{T}_{bath} = 0.03$, $\tilde{t} = 120$; (d) $\Delta\tilde{T}_{bath} = 0.04$, $\tilde{t} = 80$; (e) $\Delta\tilde{T}_{bath} = 0.05$, $\tilde{t} = 40$; The isolines of the dimensionless concentration ( $\tilde{c}$ ) were plotted with intervals: (a) $\Delta\tilde{c} = 0.005$, (b),(c) $\Delta\tilde{c} = 0.010$, and (d),(e) $\Delta\tilde{c} = 0.020$, respectively.

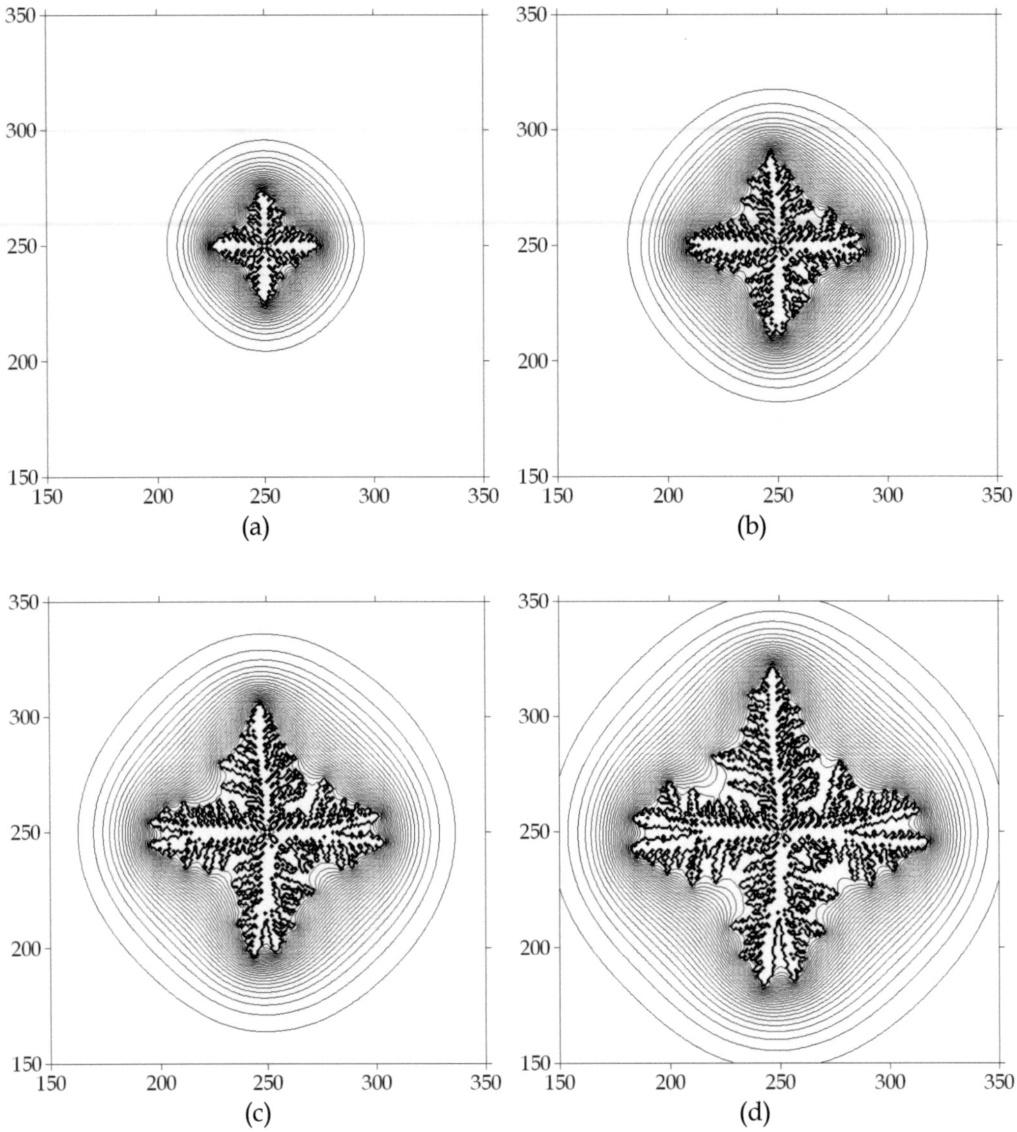


Fig. 3. Morphology of dissipative structures are formed during *Fe-B* melt crystallization and concentration fields (thin lines) in the system in some successive moments of the time $\tilde{t}$ for value of bath supercooling $\Delta\tilde{T}_{bath} = 0.055$: (a) $\tilde{t} = 5$, (b) $\tilde{t} = 10$, (c) $\tilde{t} = 15$, and (d) $\tilde{t} = 20$. The isolines of the dimensionless concentration ($\tilde{c}$) were plotted with $\Delta\tilde{c} = 0.02$ intervals.

supercooling $\Delta\tilde{T}_{bath}$ are shown in Fig. 2-5. Analysis of the evolution of dissipative structures change with supercooling of the melt revealed three types of morphology. At low supercooling of the melt (0.01-0.05) formed the "classic" dendrites, whose form is determined by the anisotropy of the kinetic coefficient (Fig. 2). As the figure shows, the distance between the secondary branches of dendrites decreases with increasing supercooling of the melt.

Fig. 4. Morphology of dissipative structures are formed during *Fe-B* melt crystallization and concentration fields (thin lines) in the system in some moments of the time $\tilde{t}$ for various values of bath supercooling $\Delta\tilde{T}_{bath}$: (a) $\Delta\tilde{T}_{bath} = 0.055$, $\tilde{t} = 75$; (b) $\Delta\tilde{T}_{bath} = 0.065$, $\tilde{t} = 75$; (c) $\Delta\tilde{T}_{bath} = 0.080$, $\tilde{t} = 67$; and (d) $\Delta\tilde{T}_{bath} = 0.100$, $\tilde{t} = 62$. The isolines of the dimensionless concentration ($\tilde{c}$) were plotted with intervals: (a) $\Delta\tilde{c} = 0.02$, (b) $\Delta\tilde{c} = 0.05$, (c) $\Delta\tilde{c} = 0.10$, and (d) $\Delta\tilde{c} = 0.05$, respectively.

However, at a further increase in supercooling of the melt a classic form of the dendrite tip becomes unstable. The tip of the dendrite splits {bifurcates). Appears branching the tips of the dendrite (Fig. 3). With further increase of supercooling of the melt (starting with $\Delta\tilde{T}_{bath} >$ 0.050) morphology of dissipative structures acquire a fractal character (formed the "fractal" dendrites, Fig. 4), which gradually evolving into a globular forms (formed "spherulitic" dendrites, Fig. 5).

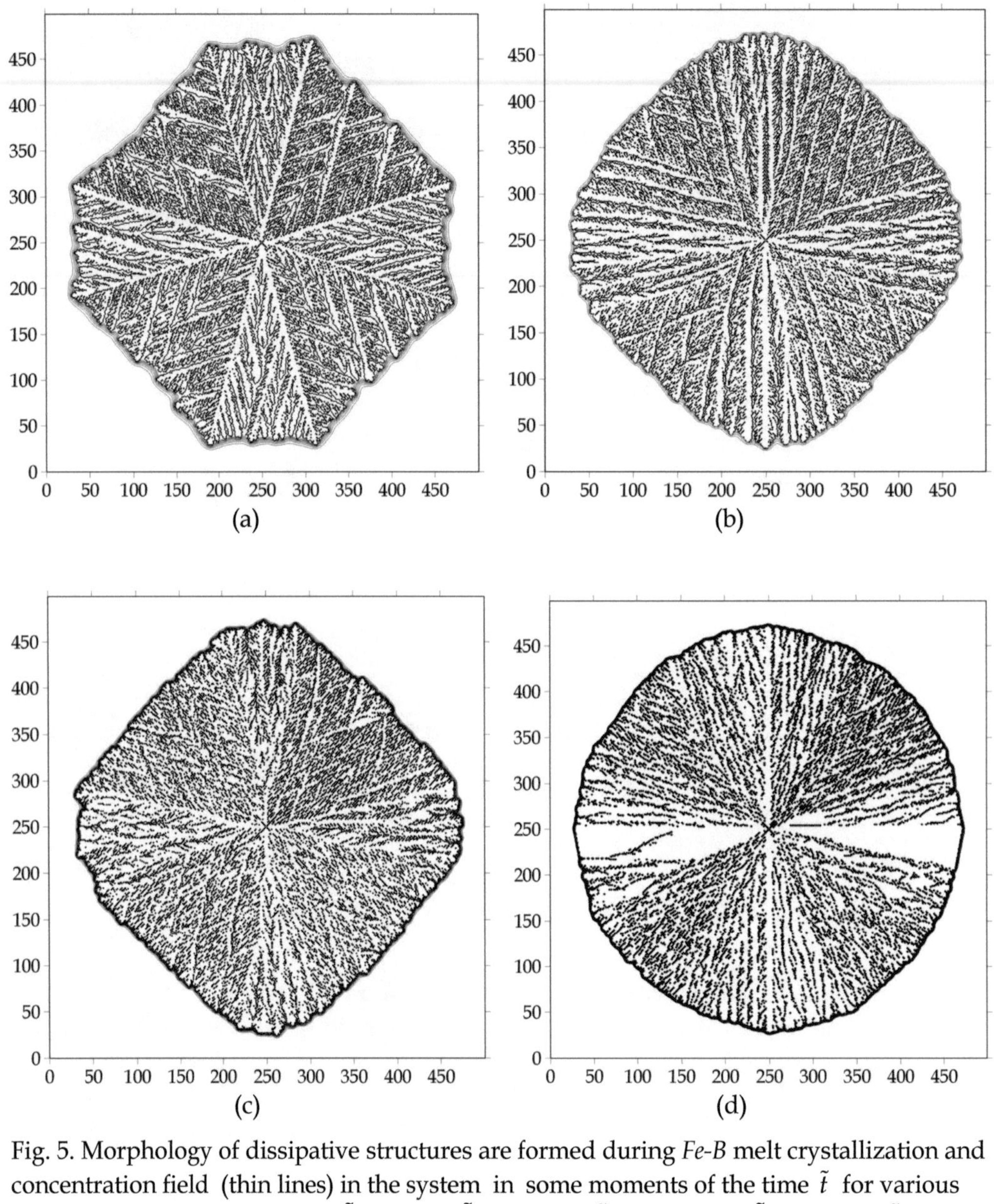


Fig. 5. Morphology of dissipative structures are formed during *Fe-B* melt crystallization and concentration field (thin lines) in the system in some moments of the time $\tilde{t}$ for various values of bath supercooling $\Delta\tilde{T}_{bath}$: (a) $\Delta\tilde{T}_{bath} = 0.200$, $\tilde{t} = 5.80$; (b) $\Delta\tilde{T}_{bath} = 0.280$, $\tilde{t} = 4.00$; (c) $\Delta\tilde{T}_{bath} = 0.430$, $\tilde{t} = 2.40$; and (d) $\Delta\tilde{T}_{bath} = 0.535$, $\tilde{t} = 2.20$. The isolines of the dimensionless concentration ($\tilde{c}$) were plotted with intervals: (a) $\Delta\tilde{c} = 0.20$, (b) $\Delta\tilde{c} = 0.20$, (c) $\Delta\tilde{c} = 0.55$, and (d) $\Delta\tilde{c} = 0.50$, respectively.

The figure 3 shows the initial stage of crystal growth. Are seen differences of the concentration fields of the solute in the melt near the sharp and in the split tops of the dendrite. Growth rate of acute vertices of the dendrite is greater than its bifurcated tops. During further growth of the dendrite is a continuous branching its tops and trunks.

These figures show that with increasing supercooling of the melt increases the degree of branching of dendrites and decreases the length of the diffusion field the concentration of solute component in the melt.

At supercooling of the melt in which the formation of globular growth forms ("spherulitic" dendrite) the region of the diffusion changes in the concentration of the dissolved component is almost completely (entirely) are localized within the macroscopic solidification front (not beyond the radius of the "spherulitic" dendrite).

With the growth process development of the solid phase (with increasing the radius of "spherulitic" dendrite) decreases the curvature of the macroscopic surface of the crystallization front. This leads to the loss of morphological stability of the dissipative structure, to reduced the overall rate of phase transformation in terms of diffusion-limited growth of the solid phase and is responsible for the transition to a kinetic regime of growth under condition when the crystal growth controlled by processes of heat transfer in a system.

Observed in computer modeling of the evolution of the morphology of dissipative structures, formed during the solidification of a supercooled melts, is in full agreement with the paradigm of self-organized criticality, describing the general laws of the development of nonequilibrium, dynamic nonlinear systems. Its essence is that with the development of a nonlinear system, it will inevitably is approaching to the bifurcation point, its stability decreases, and there are conditions under which a small push can cause an avalanche.

The development of dissipative structures is a manifestation of self-organization in nonequilibrium nonlinear systems, providing a maximum rate of increase of entropy during the phase transformation (the principle of maximum rate of entropy increase). At the crystallization of supercooled binary melt the kinetic of phase transformation in the system is controlled by transport processes of the mass and heat (removal of solute and of heat that are released during the phase transformation at the interface), whose transport coefficients are (coefficients of diffusion and thermal conductivity) differ by three orders of magnitude. This causes the appearance of two cycles in the evolution of the morphology of dissipative structures and the kinetics of phase transformation with increasing supercooling of the melt.

### 3.2 Morphology of dissipative structures formed during the solidification of a supercooled melt under conditions when the crystal growth controlled by processes of heat and mass transfer in a system

We consider the melt solidification from one centre in a system of 250 × 250 cells with the following calculation parameters: time step $\Delta\tilde{t} = (\upsilon_0^2/\alpha)\,\Delta t$ = 0.025 and spatial step $\Delta\tilde{h} = (\upsilon_0 / \alpha)\Delta h$ = 0.50. In special case the simulation is carried out for the time step $\Delta\tilde{t}$ = 0.0125 and grid spacing $\Delta\tilde{h}$ = 0.25 in a 1200 × 1200 system. The computer simulation is realized for various initial and boundary conditions: the system is at the given supercooling $\Delta T = \Delta T_{bath}$ with adiabatic boundary and the temperature on system boundary decreases from some initial value $T_{init} < T_E$ to $T = T_B$ with the given rate of melt cooling $R$.

In Fig. 6 show the morphology of the growing crystal and the temperature field (in relative magnitudes *TC/Q*) in the system in certain moments of the time for various values of the bath supercooling. With increase of the bath supercooling the morphology of growth of crystal is changed from the globular form of diffusion-limited growth (Fig. 6 a) to the cellular-dendritic form (Fig. 6 b, c) and, then, to the needle-like (Fig. 6 d) and globular (Fig. 6 e) forms of thermally controlled growth. The change of the morphology from the dendrite with a cellular lateral surface to the needle-like dendrite is shown in Fig. 6 (c).

The change in the crystal growth regimes is illustrated by the corresponding changes in the temperature field configuration. The temperature field (Fig. 6 c, d) indicates that the dendritic growth occurs in a thermally controlled regime. The isotherms are distorted under influence of the crystallization heat releasing. In the melt far from the surface of growing crystal the temperature field is concentric isolines. In the diffusion mode the globular growth form is controlled by mass-transfer of solute in the liquid at the interface. With increase of solidification velocity (with increase of supercooling) the solute trapping increases and thus the role of diffusion as limiting factor decreases (Fig. 6 a, b). Whereas the role of the heat transport as limiting factor raises, and it is the most essential at dendrite tip (Fig. 6 c,d), and at the globular form of the growth (Fig. 6 e). The dendrite - globule morphological transition takes place at the deep supercoolings $\Delta\tilde{T} > \Delta\tilde{T}^{**}$ for which the growth velocity $\tilde{V}$ in a minimum of kinetic coefficient ($\tilde{V}^{-}$ in Fig. 1) equals to the diffusion rate $\tilde{V}_D$. Although the kinetic coefficient is small in this case (i.e., exchange atomic fluxes through the interface are small), the deviation from equilibrium is large, and therefore the globule growth occurs with a high velocity and it is controlled by the removal of the released heat of solidification. The high density of isotherms in Fig. 6 e signifies that the rate of latent heat release at the all surface of crystal during solidification is large. If the rate of latent heat release is greater than the heat removal rate then the temperature increases. The temperature increase leads to a solidification velocity increase at the melt supercoolings $\Delta T > \Delta T_{Vmax}$. Thus the globular growth form is established when the heat realize and removal rates are equal.

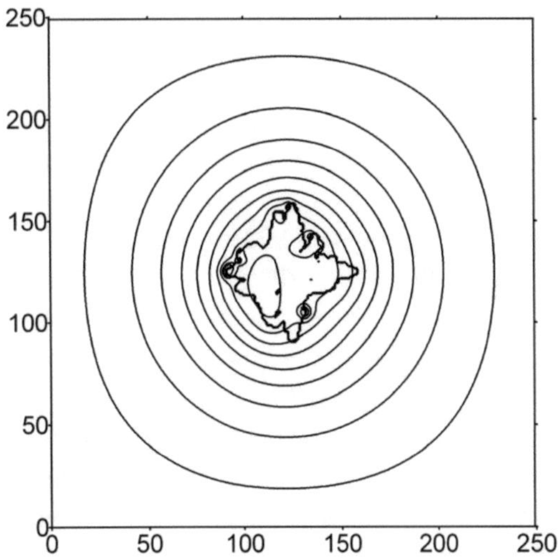


(a)

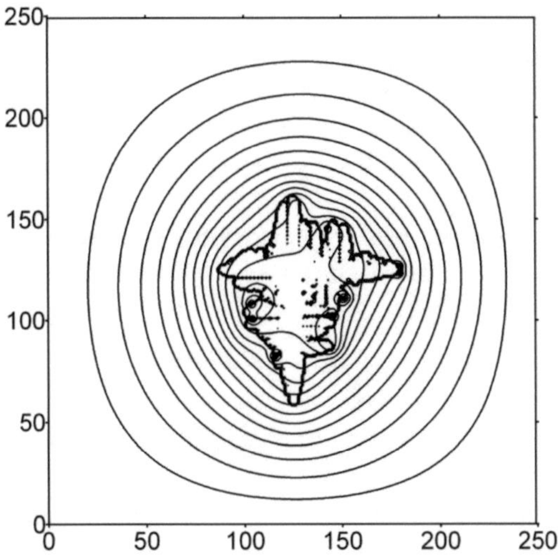


(b)

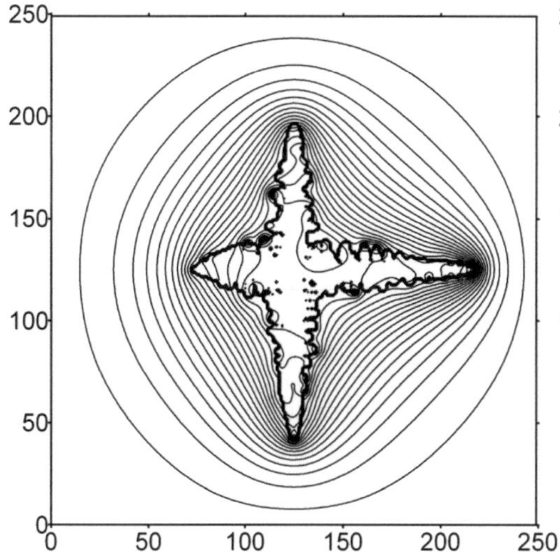


(c)

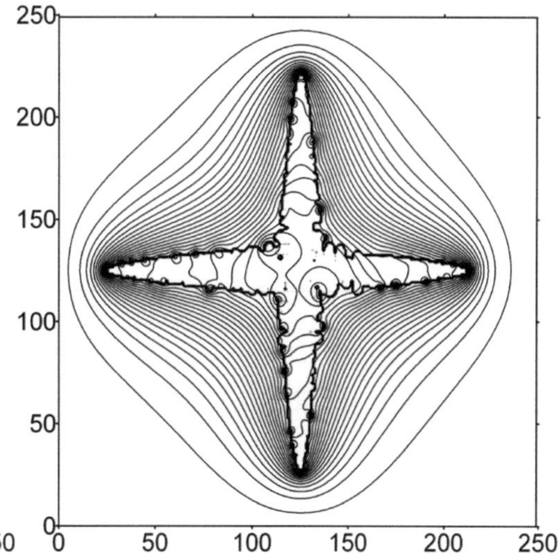


(d)

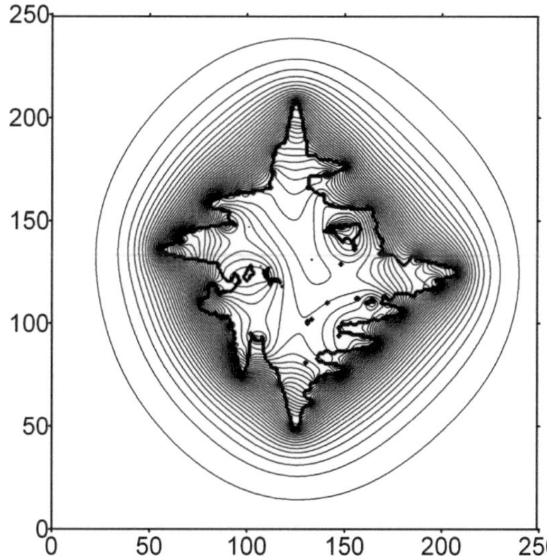


(e)

Fig. 6. Morphology of dissipative structures (bold lines) are formed during *Fe-B* melt crystallization and temperature fields (thin lines) in the system in some moments of the time $\tilde{t}$ for various values of bath supercooling $\Delta\tilde{T}_{bath}$: (a) $\Delta\tilde{T}_{bath} = 0.55$, $\tilde{t} = 500$; (b) $\Delta\tilde{T}_{bath} = 0.70$, $\tilde{t} = 500$; (c) $\Delta\tilde{T}_{bath} = 0.75$, $\tilde{t} = 250$; (d) $\Delta\tilde{T}_{bath} = 0.85$, $\tilde{t} = 175$; (e) $\Delta\tilde{T}_{bath} = 1.15$, $\tilde{t} = 250$. The isolines of the dimensionless temperature ($TC/Q$) were plotted with 0.02 intervals.

Development of instability of interface depends on the local conditions and the influence of temperature and concentration fields of branches in the nonlinear system with feedbacks. In consequence of mutual influence of the temperature and concentration fields of separate branches one of them grow with the deceleration, other is accelerating and thus a selection of the spatial period of the forming structure occurs. The scale of mutual influence of branches essentially changes at the transition from diffusion-controlled to thermally controlled mode of the growth. The sharp increase of the growth velocity of one of branches of the structure formed at the diffusion mode leads to strong distortion of a configuration of a temperature field and as a consequence to chaotic dendritic pattern. It is impossible to exclude completely and the computational grid influence. As the scales of the transfer processes of heat and mass essentially differ, and therefore when the crystallization velocity exceeds the speed of diffusion through border between the neighboring cells ($V \geq D/\Delta h$) the width of a diffusion layer $l_D$ becomes equal to the spatial step of a computational grid ($l_D = \Delta h$).

Morphology of the crystal-melt interface in subsequent moments of the time and the temperature field in the 1200 × 1200 system with the time step $\Delta\tilde{t} = 0.0125$ and grid spacing $\Delta\tilde{h} = 0.25$ at $\Delta\tilde{T}_{bath} = 0.9$ are shown in Fig. 7.

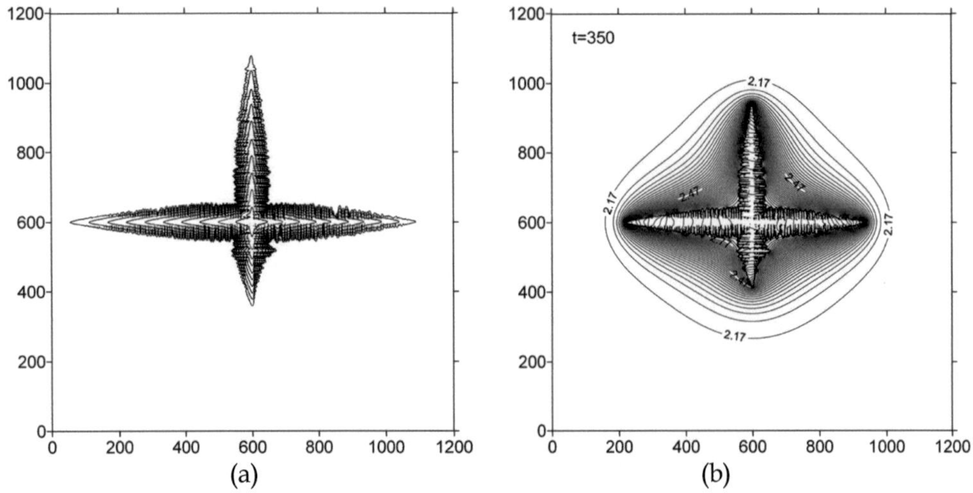


Fig. 7. Morphology of the crystal-melt interface in subsequent moments of the time in a system 1200 × 1200 cells: (a) $\Delta\tilde{T}_{bath} = 0.9$, $\tilde{t} = 50, 100, \ldots, 500$. Temperature field in the system: (b) $\Delta\tilde{T}_{bath} = 0.9$, $\tilde{t} = 350$.

The increase of the system size and the decrease of the grid step give more accurate pattern of side branches far from dendrite tip. At an initial stage of solidification of system all four branches of dendrite grow in a rapid thermal mode. The high rate of latent heat release and the small size of a crystal lead to thermal interaction between branches. As a result of this interaction the growth of one of dendrite branches is decelerated at the moment of time t > 100. The transition to cellular-dendrite growth occurs.

To study the dynamic behavior of interface during the evolution of the *Fe-B* system the change of the crystallization velocity $\tilde{V}$ and supercooling $\Delta\tilde{T}$ at the dendrite tip is calculated. Trajectories of the dendrite tip in the space of variables $\tilde{V}$ and $\Delta\tilde{T}$ are shown in Fig. 8. That is to say this is a phase portrait of a dendrite. Each point corresponds to the state of the interface

( $\tilde{V}$ , $\Delta\tilde{T}$ ) in a cell that is on the path of movement of the dendrite tip. The points are obtained by averaging of a supercooling and growth rate over the number of temporary steps, for which the two-phase cell containing the dendrite tip becomes completely solidified. The value of the supercooling $\Delta T$, for which $V = V_D$ in a maximum of kinetic coefficient, is designated as $\Delta T^*$ ($\Delta T^*$ = 303 K or $\Delta\tilde{T}^*$ = 0.499). Results of computer simulation obtained both under conditions of melt cooling on the system boundaries and under adiabatic boundary conditions show that the $V$ versus $\Delta T$ curve has an S-like character. The hysteresis characteristic for transition of such type is observed. The bottom branch of solutions is the diffusion mode and the top branch of solutions is the thermal mode. The middle branch of solutions corresponds to cellular or cellular-dendritic growth morphology. This branch corresponds to some intermediate state when the diffusion and thermal modes take place.

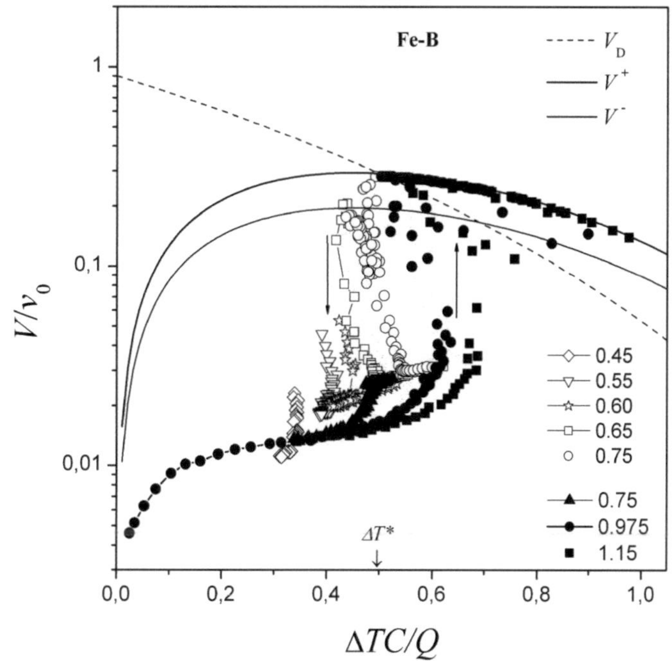


Fig. 8. Dependences of the growth velocity $\tilde{V}$ on supercooling $\Delta\tilde{T}$ of the *Fe-B* melt at dendrite tip. $\Delta T^*$ = 303 K. Solid lines: velocity of crystallization in kinetic regime. Solid points: results of simulation obtained under melt cooling on the system boundary with the rate $\tilde{R}$ = 0.001 until $\Delta\tilde{T} = \Delta\tilde{T}_B$. Open points: data of computer simulation of system with initial total supercooling $\Delta\tilde{T} = \Delta\tilde{T}_{bath}$. Values of the total $\Delta\tilde{T}_{bath}$ and final $\Delta\tilde{T}_B$ supercoolings are shown in a legend. Arrows show direction of trajectories of points (in space of variables $\tilde{V}$, $\Delta\tilde{T}$ ) during dendrite growth.

The growth of crystal at supercooling $\Delta T < \Delta T^*$ is limited by the diffusion of the dissolved component which is rejected by the interface because of a significant separation effect, since the crystallization velocity $V$ is lower than the maximum possible velocity in the kinetic regime $V < V_D$ and, therefore, the partition coefficient $k$ ($V$) weakly differs from the equilibrium value, which is much smaller than unity. The velocity jump at a dendrite tip supercooling $\Delta T \geq \Delta T^*$ corresponds to the morphological transition conditioned by the change from a diffusion to a kinetic growth regime controlled by heat transfer in the system. The segregation of dissolved component at the dendrite tip practically is absent, since the partition coefficient $k(V)$ is close to unity.

Results of computer simulation with the parameters corresponded approximately to the *Ni-B* system and experimental data on rapid alloy solidification (Eckler at al., 1992) are shown in Fig. 9. To compare our results of computer modeling to experimental data we used values: $\upsilon_0$ = 100 m/s at $\beta_0$ = 0.21 m/s K, $E_a/RT_E$ = 5.55, and $Q/C$ = 472.65 K, $\Theta$ = 3.651, $Q/RT_E$ = 1.2, $k_0$ = 0.015 taken from (Eckler at al., 1992). In the experiment the solidification occurs from multitude of the nucleus therefore crystals because of mutual influence morphologically are not developed, but the data on growth velocity specify the onset of the transition to the partitionless regime. The obtained value of a critical supercooling $\Delta T^*$ = 244 K on an order of magnitude will be agreed with the experimental data, in particular, for an alloy *Ni* - 1 at % *B* it has been established, that the growth velocity sharply increases and also solidification becomes almost partitionless at critical supercooling $\Delta T^*$ = 267 K.

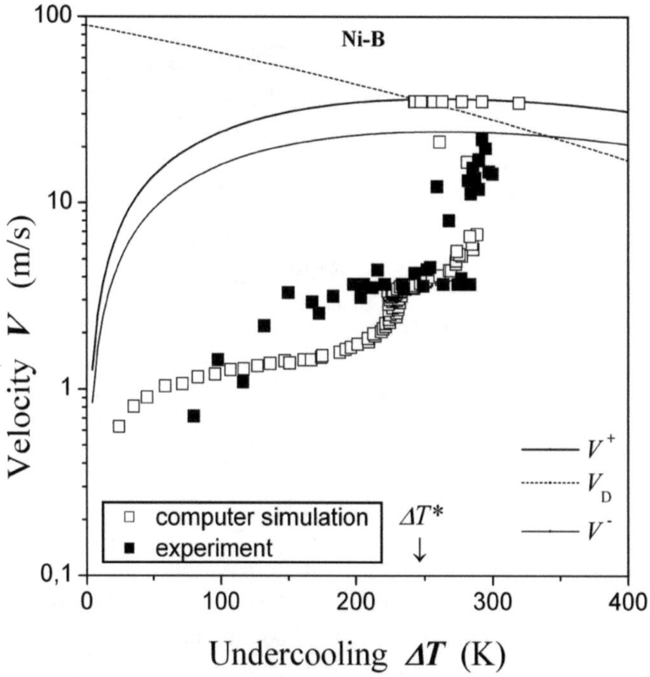


Fig. 9. Dependences of the growth velocity on supercooling of the *Ni-B* melt at dendrite tip. $\Delta T^*$ = 244 K. Solid lines: velocity of crystallization in kinetic regime. Open points: results of simulation obtained under melt cooling on the system boundary with the rate $\tilde{R}$ = 0.001 until $\Delta\tilde{T} = \Delta\tilde{T}_B$=0.9; solid points: data of experiment for *Ni–B*.

## 4. Conclusion

We have proposed a computer model that takes into account a temperature dependence of diffusion coefficient and a nonequilibrium partition of dissolved component of the alloy. In this model the dynamics of the formation of dendritic patterns from a crystallization centre has been investigated. The dependence of interface velocity $V$ on an supercooling $\Delta T$ at the dendrite tip is obtained during rapid solidification of *Fe-B* and *Ni-B* systems. The morphological transition which is conditioned by change of a diffusion growth mode on thermal growth (dendrites have the form as a needle) at some supercooling at a dendrite tip $\Delta T \geq \Delta T^*$ is detected. The $V$ versus $\Delta T$ curve has an S-like character as well as was shown for flat front of crystallization (Galenko & Danilov, 2000) and for parabolic shape of dendrite (Eckler at al., 1994) by analytic methods. Values of a critical supercooling $\Delta T^*$ and a growth velocity discontinuity depend on both the anisotropy of the kinetic coefficient, and the difference in the activation energies for atomic kinetics and for diffusion at the interface. The obtained values of a critical supercooling $\Delta T^*$ and a growth velocity discontinuity on an order of magnitude agree with well-known experimental data (Eckler at al., 1992).

It is necessary to note, that in the experiment the bath supercooling is measured, and in the present work it is investigated the change of velocity and supercooling at a dendrite tip during the crystallization at the given bath supercooling or the cooling rate. Experimental data for *Ni-B* (Eckler at al., 1992) and recent data for *Ti-Al* (Hartmann at al., 2008) testify that velocity increases with supercooling at a thermal mode. It means, that $\Delta T^* < \Delta T_{Vmax}$. In approach of the given model we obtain for *Ni-B* $\Delta T^* = 244$ K, $\Delta T_{Vmax} = 263$ K and for *Ti-Al* $\Delta T^* = 150$ K, $\Delta T_{Vmax} = 266$ K using material parameters from (Eckler at al., 1992) and (Hartmann at al., 2008), respectively. For more correct comparison with experimental data it is necessary to carry out special modelling taking into account all details of experiment. It was not the purpose of given article.

The proposed computer model allows investigate the solidification of metastable melt at the temperatures in the wide range between the equilibrium liquidus and the glass transition. As it has been noted in (Tarabaev & Esin, 2007), that for enough large rates of cooling the transition to the thermal mode can not be realized, i.e. the system becomes "frozen": when crystal growth is decelerated because recalescence is suppressed and the melt is amorphized (glass transition temperature for the *Fe-B* system is $T_G \sim 0.5\ T_M$ or the supercooling is $\Delta T_G \sim 1.5\ Q/C$ (Elliot, 1983). The crystal growth models with a collision-limited interfacial kinetics are not suitable for the description of alloy solidification at a very large supercooling when a glass formation occurs (Greer, 2001).

## 5. References


Ahmad, N. A., Wheeler, A. A., Boettinger, W. J., & McFadden, G. B. (1998). Solute trapping and solute drag in a phase-field model of rapid solidification. *Physical Review* E, Vol.58, No.3, pp. 3436-3450, ISSN 1539-3755

Aziz, M. J., & Kaplan, T. (1988). Continuous growth model for interface motion during alloy solidification. *Acta Metallurgica,* Vol.36, pp. 2335-2347, ISSN 0001-6160

Aziz, M. J. (1994). Nonequilibrium Interface Kinetics During Rapid Solidification. *Materials Science and Engineering* A, Vol.178, pp. 167-170, ISSN 0921-5093

Bartel, J., Buhrig, E., Hain, K., & Kuchar, L. (1983). *Kristallisation aus Schmelzen: A Handbook,* K. Hain, E. Buhrig, (Eds.), VEB Deutscher Verlag für Grundstoffindustrie, Leipzig

Chernov, A. A., & Lewis, J. (1967). Computer model of crystallization of binary systems: kinetic phase transitions. *Journal of Physics and Chemistry of Solids,* Vol.28, No.11, pp. 2185-2198, ISSN 0022-3697

Chernov, A. A. (1980). Crystallization processes. In: *Modern Crystallograph,* B.K. Vainshtein, (Ed.), Vol.3, pp. 7-232, Nauka, Moscow, USSR, Russian Federation

Eckler, K., Cochrane, R. F., Herlach, D. M., Feuerbacher, B., & Jurisch, M. (1992). Evidence for a Transition from Diffusion-Controlled to Thermally Controlled Solidification in Metallic Alloys. *Physical Review* B, Vol.45, pp. 5019-5022, ISSN 1098-0121

Eckler, K., Herlach, D. M., & Aziz, M. J. (1994). Search for a Solute-Drag Effect in Dendritic Solidification. *Acta Metallurgica et Materialia,* Vol.42, pp. 975-979, ISSN 0956-7151

Elliot, R. (1983). *Eutectic Solidification Processing: crystalline and glassy alloys.* Butterworths, London, Boston, ISBN 0-408-107146.

Greer, A. L. (2001). From metallic glasses to nanocrystalline solids. *Proc. of 22nd Risø Int. Symp. on Materials Science: Science of Metastable and Nanocrystalline Alloys Structure, Properties and Modelling (Risø National Laboratory Roskilde Denmark 2001),* pp. 461-466

Galenko, P. K., & Danilov, D. A. (2000). Selection of the dynamically stable regime of rapid solidification front motion in an isothermal binary alloy. *Journal of Crystal Growth,* Vol.216, No.1-4, pp. 512–526, ISSN 0022-0248

Hansen, M., & Anderko, K. (1958). *Constitution of Binary Alloys.* McGraw-Hill Book Company, INC, New York, Toronto, London

Hartmann, H., Galenko, P. K., Holland-Moritz, D., Kolbe, M., Herlach, D. M., & Shuleshova, O. (2008). Nonequilibrium solidification in undercooled Ti[sub 45]Al[sub 55] melts. *Journal of Applied Physics,* Vol.103, No.7, pp. 073509-073518, ISSN 0021-8979

Kittl, J. A., Sanders, P. G., Aziz, M. J., Brunco, D. P., & Thompson, M. O. (2000). Complete Experimental Test for Kinetic Models of Rapid Alloy Solidification. *Acta Materialia,* Vol.48, pp. 4797-4811, ISSN 1359-6454

Miroshnichenko, I. S. (1982). *Quenching from the liquid state,* Metallurgiya, Moscow, USSR

Nikonona, V. V., & Temkin, D. E. (1966). Dendrite growth kinetics in some binary melts, In: *Growth and Imperfections of Metallic Crystals,* D.E.Ovsienko, (Ed.), pp. 53-59, Naukova Dumka, Kiev, USSR

Ramirez, J. C., Beckermann, C., Karma, A., & Diepers, H.-J. (2004). Phase-field modeling of binary alloy solidification with coupled heat and solute diffusion. *Physical Review* E, Vol.69, No.5, pp. 051607-051616, ISSN 1539-3755

Tarabaev, L. P., Mashikhin, A. M., & Vdovina, I. A. (1991). Computer simulation of dendritic crystal growth. *VINITI,* Moscow, No. 2915-V91

Tarabaev, L. P., Mashikhin, A. M., & Esin, V. O. (1991). Dendritic crystal growth in supercooled melt. *J. Crystal Growth,* Vol. 114, No. 4, pp. 603 – 612, ISSN 0022-0248

Tarabaev, L. P., Psakh'e, S. G., & Esin, V. O. (2000). Computer Simulation of Segregation, Plastic Deformation, and Defect Formation during Synthesis of Composite Materials. *The Physics Metals and Metallography,* Vol.89, No.3, pp. 217 – 224, ISSN 0031-918X

Tarabaev, L. P., & Esin, V. O. (2001). Formation of Dendritic Struture upon Directional Solidification of Ternary Alloys. *Russian Metallurgy (Metally),* Vol.2001, No.4, pp. 366- 372, ISSN 0036-0295

Tarabaev, L. P., & Esin, V. O. (2007). Computer Simulation of the Crystal Morphology and Growth Rate during Ultrarapid Cooling of an *Fe-B* Melt. *Russian Metallurgy (Metally),* Vol.2007, No.6, pp. 478-483, ISSN 0036-0295

Temkin, D. E. (1970). Kinetic phase transition at the phase transition in binary alloys. *Kristallografiya,* Vol.15, No.5, pp. 884-893, ISSN 0023-4761

Wheeler, A. A., Boettinger, W. J., & McFadden, G. B. (1993). Phase-field model of solute trapping during solidification. *Physical Review* E, Vol.47, No.3, pp. 1893-1909, ISSN 1539-3755

# 7

# Phase Field Modeling of Dendritic Growth and Coarsening

Zhang Yutuo
*Shenyang Ligong University, Shenyang*
*China*

## 1. Introduction

Phase field models are known to be very powerful in describing the complex pattern evolution of dendritic growth. It is a useful method for simulating microstructure evolution involving diffusion, coarsening of dendrites and the curvature and kinetic effects on the moving solid-liquid interface. Such models are efficient especially in numerical treatment because all the governing equations are written as unified equations in the whole space of the system without distinguishing the interface from the mother and the new phase, and direct tracking of the interface position is not needed during numerical calculation. In the last decade, the phase field method has been intensively studied as a model of solidification processes [1-5]. The dendritic coarsening behavior affects the distribution of length scales, microsegregation and other microstructural characteristics of the materials, all of which determine the physical and chemical properties of materials in terms of strength, ductility and corrosion resistance. Therefore, understanding coarsening and being able to study the morphology of the dendritic structure is of technological importance[6-9]. Many properties of cast materials are intimately related to the dendritic morphology that is largely set by coarsening. Even if the effect of the dendritic microstructure is altered by subsequent heat treatment, they rarely fully disappear.

In this part, the dendritic growth of and the subsequent dendritic coarsening as well as the effect of undercooling on coarsening in Al-2mol%Si alloy during isothermal solidification are simulated using phase field model.

## 2. Phase field model

The phase field model used in this paper was developed by Kim et al. So it would be briefly mentioned here. Readers can refer to literatures [10,11] for details of the formulation. The model includes two variables: one is a phase field $\phi(x,y.t)$ and the other is a concentration field $c(x,y,t)$. The variable $\phi(x,y.t)$ is an ordering parameter at the position $(x,y)$ and the time $t$, $\phi$=1 means being solid and $\phi$=0 liquid. The solid-liquid interface is expressed by the steep layer of $\phi$ connecting the value 0 and 1. So the phase field model can be described as

$$\frac{\partial \phi}{\partial t} = M\left(\varepsilon^2(\theta)\nabla^2\phi - f_\phi\right) \tag{1}$$

$$\frac{\partial c}{\partial t}=\nabla\bullet\left(\frac{D(\phi)}{f_{cc}}\nabla f_c\right) \quad (2)$$

Where, $$f(c,\phi)=h(\phi)f^{S}(c_S)+(1-h(\phi))f^{L}(c_L)+Wg(\phi) \quad (3)$$

$$D(\phi)=D_L+h(\phi)(D_S-D_L) \quad (4)$$

$$h(\phi)=\phi^3\left(6\phi^2-15\phi+10\right) \quad (5)$$

$$g(\phi)=\phi^2\left(1-\phi^2\right) \quad (6)$$

$$c=h(\phi)c_S+(1-h(\phi))c_L \quad (7)$$

$$\mu^S((c_S(x,t))=\mu^L(c_L(x,t)) \quad (8)$$

$$f^S=c_S f_B^S(T)+(1-c_S)f_A^S(T) \quad (9)$$

$$f^L=c_L f_B^L(T)+(1-c_L)f_A^L(T) \quad (10)$$

$$\varepsilon(\theta)=\varepsilon_0\left\{1+v\cos(k\theta)\right\} \quad (11)$$

$$M^{-1}=\frac{\varepsilon^2}{\sigma}\left[\frac{RT}{V_m}\frac{1-k^e}{m^e}\frac{1}{\mu}+\frac{\varepsilon}{D_L\sqrt{2W}}\varsigma\left(c_S^e,c_L^e\right)\right] \quad (12)$$

$$\varsigma\left(c_S^e,c_L^e\right)=f_{cc}^S\left(c_S^e\right)f_{cc}^L\left(c_L^e\right)\left(c_L^e-c_S^e\right)^2 \times\int_0^1\frac{h(\phi)\left[1-h(\phi)\right]d\phi}{\left[1-h(\phi)\right]f_{cc}^S\left(c_S^e\right)+h(\phi)f\phi\left(c_L^e\right)\phi(1-\phi)} \quad (13)$$

$$\varepsilon=\sqrt{\frac{6\lambda}{2.2}\sigma} \quad (14)$$

$$W=\frac{6.6\sigma}{\lambda} \quad (15)$$

Where $M$ and $\varepsilon$ are the phase field mobility and gradient energy coefficient, respectively. $f$ is the free energy density of the system. The subscripts under $f$ indicate the partial derivatives. $D(\phi)$ is the diffusivity of solute as a function of phase field. $D_S$ and $D_L$ are the diffusive coefficient in the solid and liquid respectively. $h(\phi)$ and $W$ correspond to solid fraction and the height of double-well potential, respectively. $c_L$ and $c_S$ are the solute concentration in liquid and solid, respectively. $\varepsilon_0$ is the mean value of $\varepsilon$, $\theta$ is the angle between the normal to the interface and the x-axis, $\theta$=arctan($\phi_y$/ $\phi_x$). $v$ is the strength of anisotropy and $k$ is the mode number, $k$=4. $m^e$ is equilibrium slope of the liquids, $k^e$ is the equilibrium partition coefficient. The phase field parameters of $\varepsilon$ and $W$ are related to the interface energy $\sigma$ and the interface width $2\lambda$.

In addition, stochastic noise introduced into the phase field model causes fluctuations at the solid/liquid interface that leads to the development of a dendrite structure. Herein, noise is introduced by modifying the phase field equation

$$\frac{\partial \phi}{\partial t} \to \frac{\partial \phi}{\partial t} + 16g(\phi)\chi\omega \tag{16}$$

Where $\chi$ is a random number distributed uniformly between -1 and 1, a new number is generated for every point of the grid, at each time-step. $\omega$ is an amplitude of the fluctuations, here $\omega$=0.01 in the calculation.

The Al-2mol%Si alloy is considered in this work. Isothermal computations are performed using the model described above. The grid sizes of the phase field and the concentration field are $1.0 \times 10^{-8}$ m. The governing equations are discretized on uniform grids by using an explicit finite difference scheme. The thermo-physical data given as: $\sigma$=0.093 J·m$^{-1}$, $T_m$=922K, $V_m$=1.06×10$^{-5}$ m$^3$/mole, $k^e$=0.0807, $D_L$=3×10$^{-9}$ m$^2$·s$^{-1}$, $D_S$=1×10$^{-12}$ m$^2$·s$^{-1}$, $m^e$=-939.0, $v$=0.03, $\omega$ =0.01.

## 3. Results and discussion

### 3.1 Simulation of single dendrite growth during isothermal solidification

Fig.1 shows the evolution of simulated single dendrite during isothermal solidification of Al-2mol%Si alloy at the temperature of 870K. Fig.2 illustrates the concentration profiles calculated during single dendrite isothermal solidification of Al-2mol%Si alloy at the temperature of 870K. It can be seen from Fig.1, the nucleus grows and becomes unstable and then the dendrite forms. In the early stage of solidification, the dendrite develops the main arms along the crystallographic orientations. The secondary and tertiary arms as well as the necking phenomenon can be observed. In the case of equiaxed dendrite growth, there exists a solute build-up ahead of the dendrite tip, as shown in Fig.2. The crystal shape is dictated mainly by the diffusion of solute. Nevertheless, a remaining anisotropy in properties leads to the growth of dendrite arms in specific crystallographic direction.

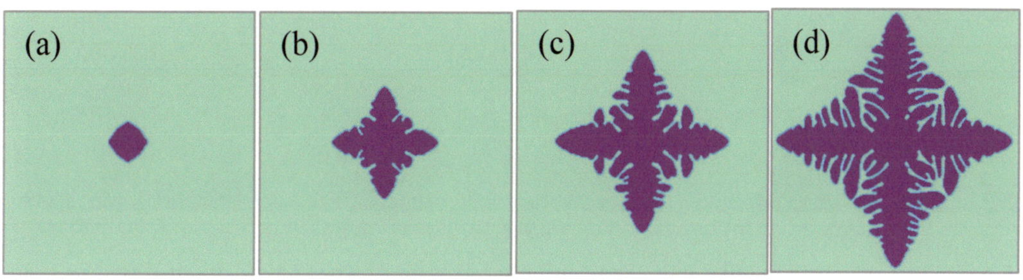


Fig. 1. Simulated single dendrite evolution during isothermal solidification of Al-2mol%Si alloy at the temperature of 870K. The snapshots (a), (b), (c) and (d) correspond to solidification time of 0.01ms, 0.04ms, 0.08ms and 0.14ms respectively.

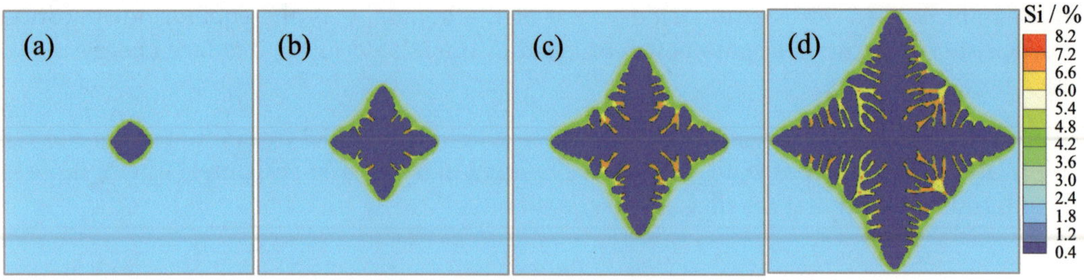


Fig. 2. Concentration profiles calculated during single dendrite isothermal solidification of Al-2mol%Si alloy at the temperature of 870K. The snapshots (a), (b), (c) and (d) correspond to solidification time of 0.01ms, 0.04ms, 0.08ms and 0.14ms respectively.

The oriented growth of single dendrite from the bottom-left corner toward the top-right corner of the square domain was simulated in Al-2mol%Si alloy. Fig.3 shows the evolution of oriented growth of single dendrite during isothermal solidification in Al-2mol%Si alloy at temperature of 870K. Fig.4 illustrates the concentration profiles during isothermal solidification of Al-2mol%Si alloy at temperature of 870K. The snapshots (a), (b), (c), (d) and (e) in Fig.3 and Fig.4 correspond to solidification time of 0.04, 0.10, 0.14, 0.18, and 0.22ms respectively. Fig.5 shows the snapshot of the oriented dendrite at solidification time of 0.26ms. Fig.6 shows the curve of dendritic tip velocity versus time.

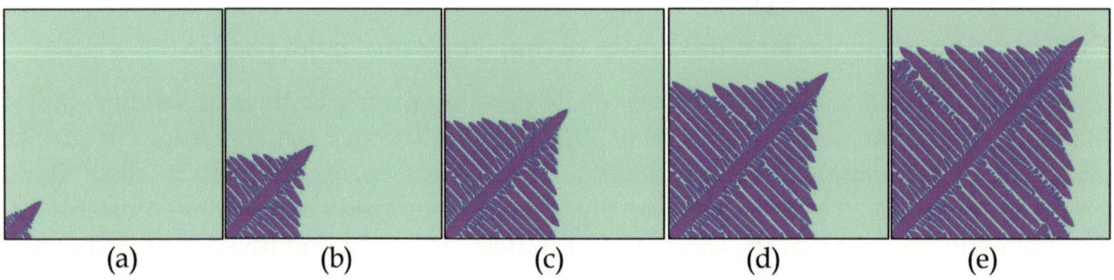


Fig. 3. Simulated oriented growth of single dendrite during isothermal solidification in Al-2mol%Si alloy at temperature of 870K. The snapshots (a), (b), (c), (d) and (e) correspond to solidification time of 0.04, 0.10, 0.14, 0.18 and 0.22ms respectively.

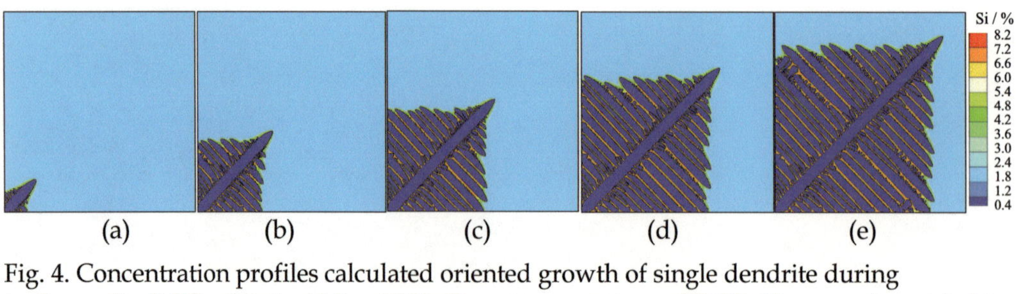


Fig. 4. Concentration profiles calculated oriented growth of single dendrite during isothermal solidification in Al-2mol%Si alloy at temperature of 870K. The snapshots (a), (b), (c), (d) and (e) correspond to solidification time of 0.04, 0.10, 0.14, 0.18, and 0.22ms respectively.

The simulation results show the morphology of dendrite with primary and secondary arms as well as tertiary arms. The process of dendrite growth and the competition between the dendrite arms are reproduced. At the early stage of solidification, the dendrite develops the main arms along the crystallographic orientations. Near the primary arm, the small secondary arms compete with each other and some overgrown secondary arms survive. The dendrite grows fast at the early stage as shown in Fig.6, in which the dendritic tip velocity increases steeply with the time. In the process of dendrite growth, the secondary arms are sometimes eliminated by their neighbors, and a number of them grow perpendicularly to the primary arm. The survived secondary arms grow until being screened by the tertiary arms, whereas the dendritic tip velocity changes in a small range due to the addition of noise.

Because of the concentration redistribution in the front of solid-liquid interface, the interdendritic liquid always has a different composition compared to that of the dendrite arms. The concentration of Si in dendrite arms is the lowest. The highest concentration corresponds to the interdendritic liquid, as shown in Fig.4.

In addition, an interesting phenomenon is found that the tertiary arms only grow at one side of some secondary arms which can be seen in Fig.5. The simulated result is in accordance with the result of Seong Gyoon KIM[12].

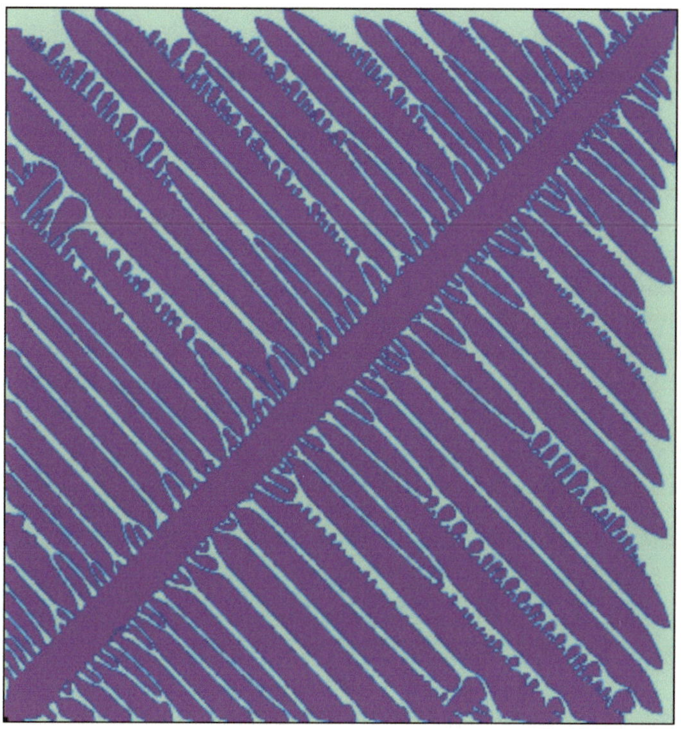

Fig. 5. Snapshot of oriented growth dendrite at time of 0.26ms

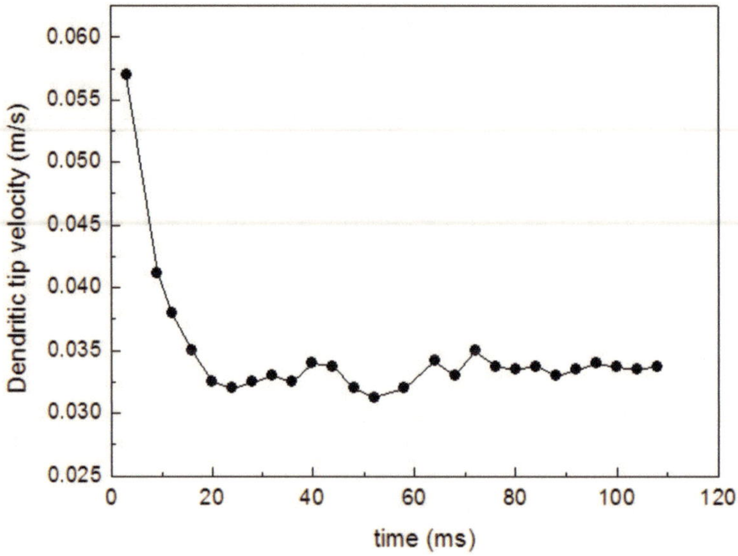


Fig. 6. The curve of growth velocity versus time

### 3.2 Simulation of multi-dendrite growth during isothermal solidification

#### 3.2.1 Simulation of multi-dendrite growth during isothermal solidification

Fig.7 shows the evolution of simulated multi-dendritic growth during isothermal solidification of Al-2mol%Si alloy at the temperature of 880K. Fig.8 illustrates the concentration profiles calculated during isothermal solidification of Al-2mol%Si alloy at the temperature of 880K. The dendrites grow freely and independently in the melt and finally impinge on one another for arbitrarily oriented crystals, which can be seen from Fig.7 (a) and (b). In the process of dendritic growth, the secondary arms are eliminated by their neighbors, and a number of them grow perpendicularly to the primary arm. However, the growth of some main arms is suppressed by nearby dendrite. As solidification proceeds, growing and coarsening of the primary arms occur, together with the branching and coarsening of the secondary arms, as shown in Fig.7 (c) to (e). Due to the concentration redistribution in the front of solid-liquid interface, as shown in Fig.8, the interdendritic liquid always has a different composition compared to that of the dendrite arms. The concentration of Si in dendrite arms is the lowest. The highest concentration corresponds to

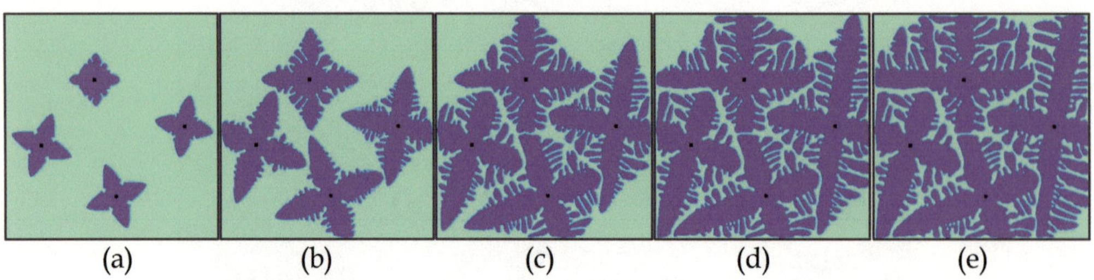


Fig. 7. Simulated multi-dendrite evolution during isothermal solidification of Al-2mol%Si alloy at the temperature of 880K. The snapshots (a), (b), (c), (d) and (e) correspond to solidification time 0.04, 0.08, 0.12, 0.16 and 0.2ms respectively.

the interdendritic liquid. Once the diffusion fields of dendrite tips come into contact with those of the branches growing from the neighboring dendrites, the dendrites stop growing and being to ripen and thicken.

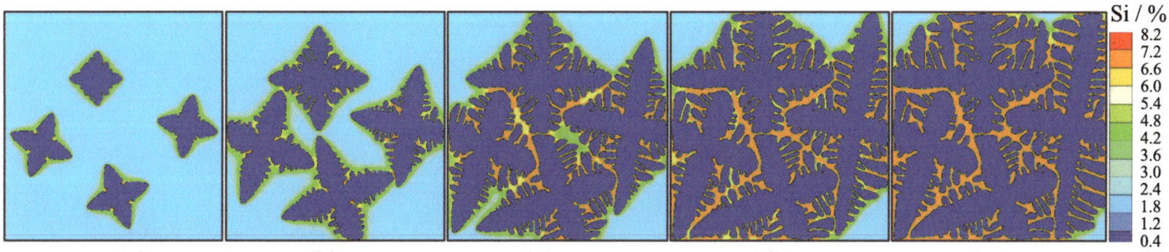


(a) (b) (c) (d) (e)

Fig. 8. Concentration profiles calculated during multi-dendrite isothermal solidification of Al-2mol%Si alloy at the temperature of 880K. The snapshots (a), (b), (c), (d) and (e) correspond to solidification time 0.04, 0.08, 0.12, 0.16 and 0.2ms respectively.

### 3.2.2 Effect of anisotropy strength on the dendrite morphology

Fig.9 shows the effect of anisotropy strength on the dendrite morphology during isothermal solidification of Al-2mol%Si alloy at 870K for solidification time of 0.04ms, 0.1ms and 0.2ms correspond to anisotropy strength 0, 0.01, 0.03 and 0.08 respectively. All the parameters except the strength of anisotropy ($v$) are fixed and $v$ is increased gradually from zero. Fig.9 (a) shows the microstructure where $v$=0, namely perfect isotropic growth. In Fig.9 (b)-(d), the strength of anisotropy ($v$) is $v$=0.01, $v$=0.03 and $v$=0.08 respectively.

In Fig.9 (a), where $v$=0, namely perfect isotropic growth is considered, the simulated patterns similar to the viscous fingering obtained. Tip splitting is seen as the dendrite growth. This figure shows the growth of a dense-branching morphology. For $v$=0.01 in Fig.9 (b), the shape of the crystal growth has both the features of isotropic and the dendrite structure. It can be seen the dendrite and the viscous finger-like structure. For $v$=0.03 in Fig.9 (c), the results show one typically dendrite structure. For $v$=0.08 in Fig.9 (d), the results show another type of dendrite structure, in which the secondary arms are in destabilization state and the smaller dendrite arms are gradually disappeared. In addition, it can be seen the "necking" during the dendritic growth process. From these simulations, the dendrites grow up faster along with the crystal-axis direction.

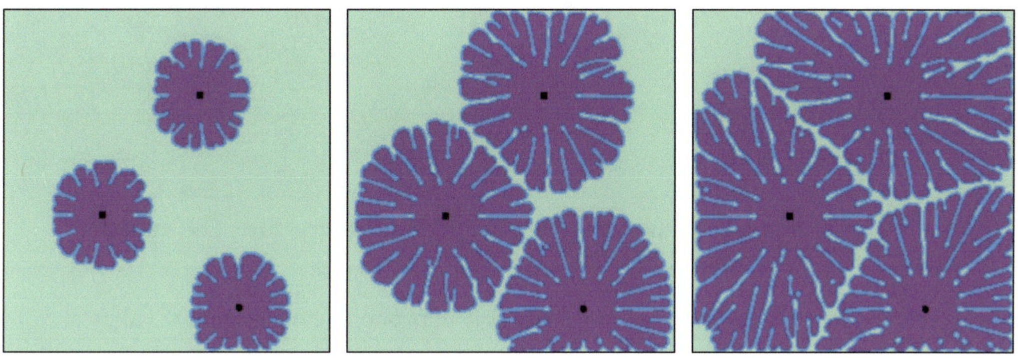

(a) For $v$=0, namely perfect isotropic growth. This figure shows the growth of a dense-branching morphology. The snapshots correspond to time of 0.04ms, 0.1ms and 0.2ms, respectively.

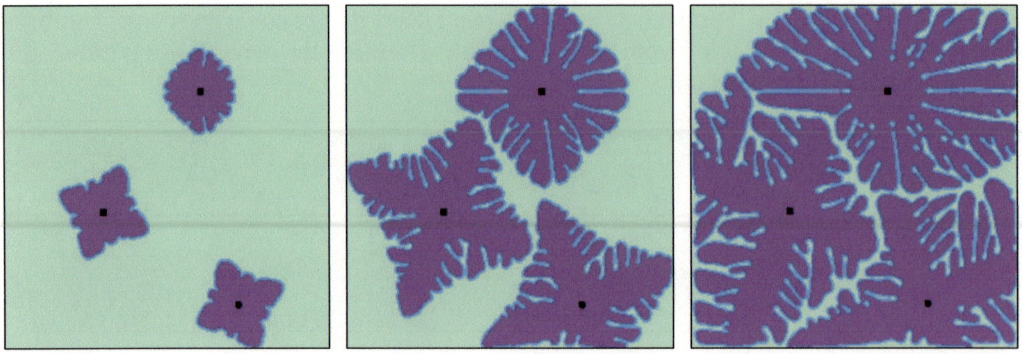

(b) For $v$=0.01, the shape of the crystal growth has both the feature of isotropic and of the dendrite structure. The snapshots correspond to time of 0.04ms, 0.1ms and 0.2ms, respectively.

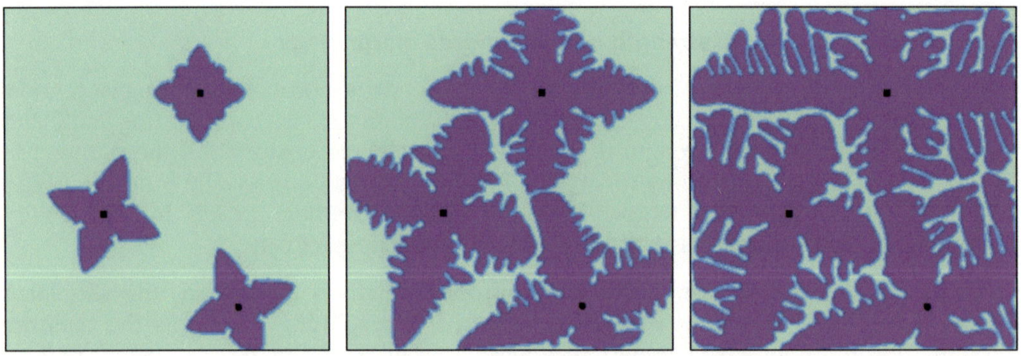

(c) For $v$=0.03, the simulation patterns of dendrite growth show typical dendrite structure. The snapshots correspond to time of 0.04ms, 0.1ms and 0.2ms, respectively.

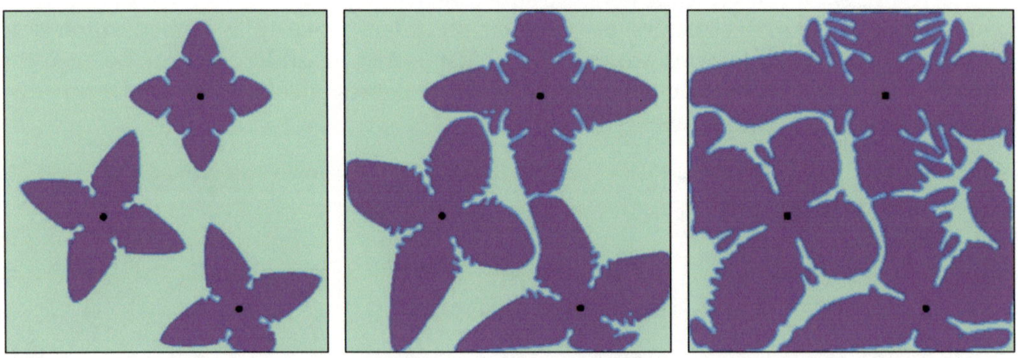

(d) For $v$=0.08, the simulations show another type of dendrite structure. The snapshots correspond to time of 0.04ms, 0.1ms and 0.2ms, respectively.

Fig. 9. Effect of anisotropy strength on the dendrite morphology of Al-2mol%Si alloy at 870K

Comparing these simulations, it is obvious that the dendrite structure is very sensitively dependent on the strength of anisotropy ($v$).

### 3.3 Dendritic coarsening under different undercoolings

Dendritic coarsening for Al-2mol%Si alloy under different undercoolings of ΔT=37K, ΔT=47K, ΔT=57K and ΔT=67K during isothermal holding is carried out. Fig.10 illustrates the fraction of solid phase as a function of solidification time for different undercoolings. The value of fraction of solid phase with solidification time under different undercoolings is listed in Table1. Fig.11 shows the evolution of the dendritic microstructure during isothermal holding at time of 0.20ms, 0.30ms, 0.40ms, 0.58ms, 0.72ms and 1.60ms respectively at undercoolings of (a) ΔT=37K, (b) ΔT=47K, (c) ΔT=57K and (d) ΔT=67K.

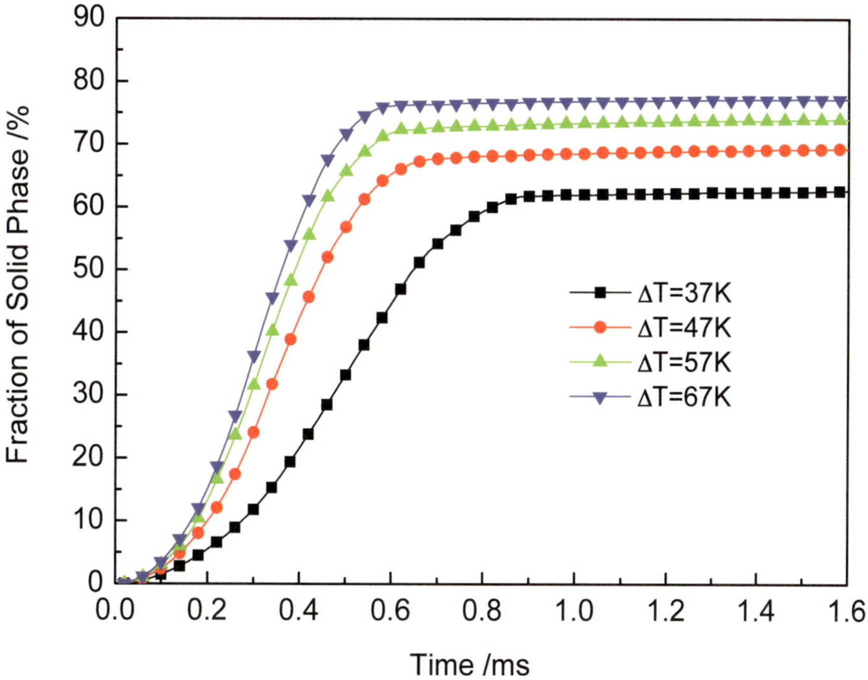


Fig. 10. The fraction of solid phase as a function of solidification time under different undercoolings

| Solidification time | Fraction of solid phase (%) | | | |
|---|---|---|---|---|
| (ms) | ΔT=37K | ΔT=47K | ΔT=57K | ΔT=67K |
| 0.20 | 5.5 | 10.0 | 13.4 | 15.2 |
| 0.30 | 11.9 | 24.0 | 31.6 | 36.4 |
| 0.40 | 21.6 | 42.5 | 52.0 | 57.7 |
| 0.58 | 42.4 | 64.2 | 71.2 | 76.0 |
| 0.72 | 55.5 | 67.8 | 72.7 | 76.4 |
| 1.60 | 62.6 | 69.2 | 73.9 | 77.2 |

Table 1. The fraction of solid phase with solidification time under different undercoolings

The dendritic morphology and dendritic coarsening mechanisms as well as the solidification kinetics are varied with the undercooling. With the undercooling increasing, the fraction of solid phase is also increased. The curve of solidification kinetics is similar for ΔT=37K, ΔT=47K, ΔT=57K and ΔT=67K, but the final fraction of solid phase is quite different, and the coarsening time is also different. The coarsening time is 0.88ms, 0.70ms, 0.62ms and 0.58ms for undercooling ΔT=37K, ΔT=47K, ΔT=57K and ΔT=67K respectively, after that the fraction of solid phase keeps stable basically, and the stable fraction of solid phase is 62.6%, 69.2%, 73.9% and 77.2% respecitviely. The higher of the undercooling, the faster of the dendrite growth and the shorter of reaching coarsening time. When undercooling ΔT=37K and ΔT=47K, the evolution of microstructure show typical dendritic morphology, as shown in Fig.11 (a) and (b). The mechanisms of isothermal dendritic coarsening are melting of samll dendrite arms (as indicated in the circled area in Fig.11 (a) and (b), coalescence of dendrites (as indicated in the round corner rectangled area in Fig.11 (a) and (b)) and smoothing of dendrites (as indicated in the rectangled area in Fig.11 (a) and (b)). It shows dendrite remelting from tips towards roots and coalescence between neighboring branches. Dendrite remelting is found to be dominant in the early stage of dendrite growth, whereas

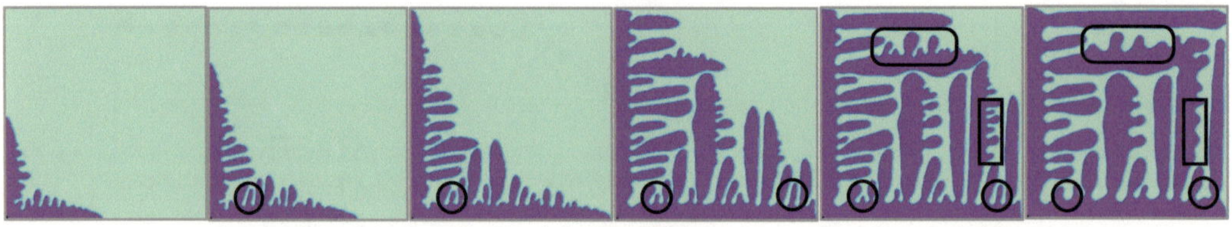

(a) ΔT=37K

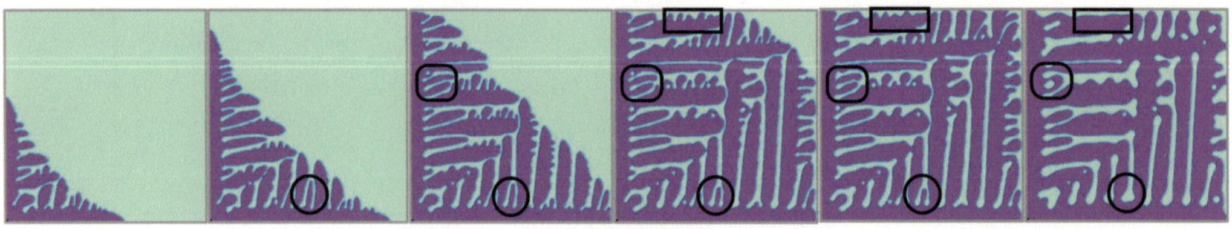

(b) ΔT=47K

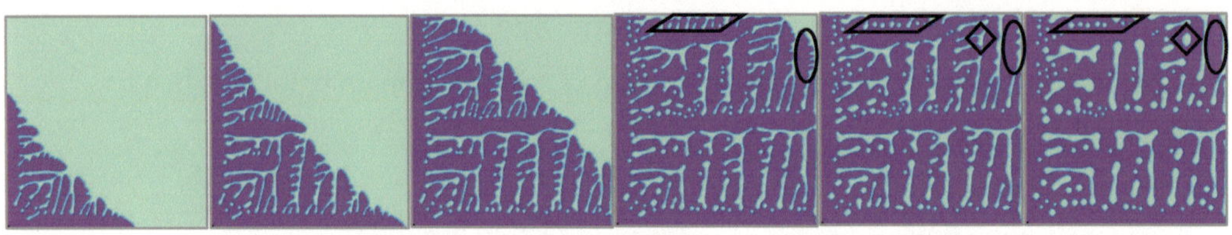

(c) ΔT=57K

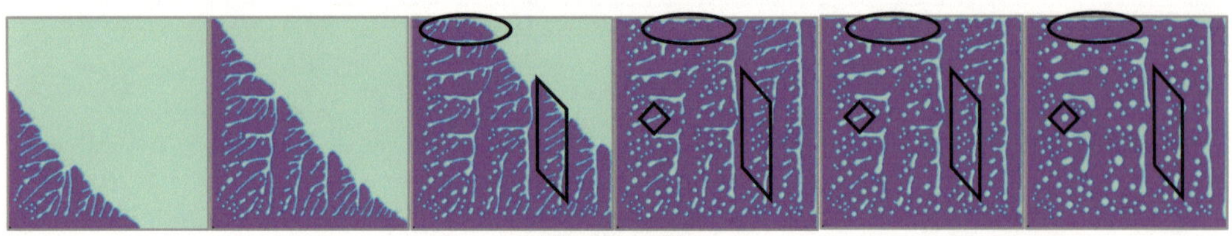

(d) ΔT=67K

Fig. 11. The evolution of the dendritic microstructure for Al-2mol%Si alloy during isothermal holding at time of 0.20ms, 0.30ms, 0.40ms, 0.58ms, 0.72ms and 1.60ms respectively for undercooling of (a) ΔT=37K, (b) ΔT=47K, (c) ΔT=57K and (d) ΔT=67K coalescence and smoothing of dendrites are dominant during isothermal holding. When undercooling ΔT=57K and ΔT=67K, the evolution of microstructure shows viscous dendritic morphology, as shown in Fig.11 (c) and (d). The mechanisms of isothermal dendritic coarsening are coalescence of dendrites with the entrapment of liquid droplets (as indicated in the parallelogram area in Fig.11 (c) and (d)) and smoothing of dendrites (as indicated in the elliptical area in Fig.11 (c) and (d)) as well as the rounding of interdendritic liquid droplets (as indicated in the rhombus area in Fig.11 (c) and (d)).

With the undercooling increasing, the interdendritic liquid is reducing. When solidification time is 1.60ms, the interdendritic liquid takes up 37.4% of the domain for ΔT=37K, and 22.8% of the domain for ΔT=67K. In addition, the pattern of the interdendritic liquid is also different with the undercooling. When the undercooling ΔT=37K, the liquid phases in the dendritic structure show platelike structure. With the undercooling increase, the liquid phases in the dendritic structure show rodlike structure. The simulated results are in accordance with Wang's simulation results [13].

## 4. Conclusions

1. The simulation of single dendrite and multi-dendrite growth for Al-2mol%Si alloy during isothermal solidification are carried out by phase field method. The primary and secondary arms as well as the necking phenomenon can be observed. For the oriented growth of single dendrite from the bottom-left corner toward the top-right corner of the square domain The survived secondary arms grow until being screened by the tertiary arms. An interesting phenomenon is found that the tertiary arms only grow at one side of some secondary arms.
2. For multi-dendrite simulation, the dendrites grow freely and independently in the melt and finally impinge on one another. As solidification proceeds, growing and coarsening of the primary arms occurs, together with the branching and coarsening of the secondary arms. Due to the concentration redistribution in the front of solid-liquid interface, the interdendritic liquid always has a different composition compared to that of the dendrite trunks. When the diffusion fields of dendrite tips come into contact with those of the branches growing from the neighboring dendrites, the dendrites stop growing and being to ripen and thicken.
3. For Al-2mol%Si alloy, when the undercooling is ΔT=37K and ΔT=47K, the evolution of microstructure shows typical dendritic morphology, and the mechanisms of isothermal dendritic coarsening are melting of samll dendrite arms, coalescence of dendrites and smoothing of dendrites. When undercooling is ΔT=57K and ΔT=67K, the evolution of microstructure shows viscous dendritic morphology, and the mechanisms of isothermal dendritic coarsening are coalescence of dendrites with the entrapment of liquid droplets and smoothing of dendrites as well as the rounding of interdendritic liquid droplets. The solidification kinetics is similar for different undercoolings, but the coarsening time and the final fraction of solid phase is quite different. The higher of the undercooling, the faster of the dendrite growth and the shorter of reaching coarsening time.

## 5. Acknowledgement

This work was supported by the Natural Science Foundation of Liaoning Province (20092061). The author would like to thank her graduate students of Sun Qiang, Cui Haixia and Hu Chunqing.

## 6. References

[1] P. Zhao, M. Venere b, J.C. Heinrich, D.R. Poirier. Modeling dendritic growth of a binary alloy. Journal of Computational Physics, Vol.188 (2003): 434-461

[2] W. J. Boettinger, J. A.Warren, C. Beckermann, and A. Karma. Phase-Field Simulation of Solidification. Annu. Rev. Mater. Res., Vol.32 (2002): 163-94

[3] D.Kammer, P.W.Voorhees. The morphological evolution of dendritic microstructures during coarsenig. Acta Materialia, Vol.54 (2006): 1549-1558

[4] Sun Qiang, Zhang Yutuo, Cui Haixia and Wang Chengzhi. Phase field modeling of multiple dendritic growth of Al-Si binary alloy under isothermal solidification. China Foundry, Vol.5 (2008): 265-267

[5] D. Danilov, B. Nestler. Phase-field simulations of solidification in binary and ternary systems using a finite element method. Journal of Crystal Growth, Vol.275 (2005): e177-e182

[6] W. J. Boettinger, J. A. Warren, C. Beckermann. Phase-field simulation of solidification. Annu. Rev. Mater. Res., Vol.32 (2002): 163-94

[7] Junjie Li, Jincheng Wang, Gencang Yang. Investigation into microsegregation during solidification of a binary alloy by phase-field simulations. Journal of Crystal Growth, 311(2009):1217-1222

[8] Ruijie Zhang, Tao Jing, Wanqi Jie, Baicheng Liu. Phase-field simulation of solidification in multicomponent alloys coupled with thermodynamic and diffusion mobility databases. Acta Materialia, 54(2006): 2235-2239

[9] B.LI, H.D.Brody and A.Kazimirov. Real time synchrotron microradiography of dendrite coarsening in Sn-13 Wt Pct Bi alloy. Metallurgical and Materials Transactions A, Vol.38 (2007): 599-605.

[10] Toshio Suzuki, Machiko Ode, Seong Gyoon Kim, Won Tae Kim, Phase-field model of dendritic growth, Journal of Crystal Growth, Vol. 237-239 (2002): 125-131

[11] Seong Gyoon Kim, Won Tae Kim, Toshio Suzuki. Phase-field model for binary alloys, Physical Review E, Vol.60 (1999): 7186-7197

[12] Seong Gyoon KIM, Won Tae KIM, Jae Sang LEE, Machiko ODE and Toshio SUZUKI. Large scale simulation of dendritic growth in pure undercooled melt by phase-field model. ISIJ International, Vol.39(1999): 335-340

[13] Jincheng Wang, Gencang Yang. Phase-field modeling of isothermal dendritic coarsening in ternary alloys. Acta Materialia, 56(2008): 4585–4592